I̲A̲

Industrial Archaeology

A SERIES EDITED BY
L. T. C. ROLT

5
Civil Engineering:
Railways

Civil Engineering: Railways

Bryan Morgan

Longman

LONGMAN GROUP LIMITED
London

*Associated companies, branches and representatives
throughout the world*

© *Bryan Morgan* 1971
First Published 1971

ISBN 0 582 12792 0

*Printed in Great Britain
by W & J Mackay & Co Ltd, Chatham*

Contents

List of Illustrations

vii

Line Drawings in the Text

Acknowledgements

In the acknowledgements to my *Railway Relics* I offered my thanks to some hundred informants. Much of their special knowledge has also been included here, and in several cases they have kindly replied to supplementary questions too. Rather than repeat a very lengthy list, may I this time simply issue a collective recognition of my indebtedness, with a special mention of the help of Dr Michael Lewis, Ian Carr, John Harris, Gordon Riddle, David Walters, and the Midland region of British Railways?

My work on the present book has been made the more pleasant and accurate through the close cooperation of my colleague, John Snell. The general editor of this series, L. T. C. Rolt, has been most helpful; and above all, I have been immensely aided by the professional services of J. Horsley Denton. All errors are, of course, my own responsibility.

Finally, I would like to take an overdue opportunity to thank Mrs Elizabeth Kerr, who has typed me through my last five books as well as numerous articles, and who has added to her skill in interpreting my handwriting the special knowledge of the wife of a civil engineer.

Photographic credits are gratefully extended to: R. Buttrell (Plates 23, 25, 27, 28, 31, 32, 33, 36, 38, 39, 40), I. S. Carr (Plates 2, 3, 4, 5, 7, 20, 35), J. Horsley Denton (Plates 24, 29, 30, 34, 44), J. Snell (Plates 1, 41); and to British Railways, Midland and Southern regions (Plates 14, 37, 43). The line drawings have been specially made for this book unless otherwise credited.

B. M.
London, 1970

Every valley shall be exalted, and every mountain
and hill shall be made low; and the crooked shall be
made straight, and the rough places plain.

ISAIAH

Introduction

The history of technology has been defined as that of the coincidence
of the need and the tool. A third and equally necessary component
for progress, however, is even today the man. Furthermore, it is an
over-simplification to see technical advance as an arbitrary series of
problems being plugged like bath-pipes with solutions of the right
shape. Rather do questions and answers, demands and responses,
follow each other in a helical cycle with the solution to one problem
itself generating the next need. A fairly simple example of this type
of inter-action can be seen in the history of firearms, where the
invention of better explosives led to the faster discharge of weapons,
which led to fouled barrels, which led to a need for better explosives. . . . And one of a dozen contemporary illustrations of this
principle can be found in the relationship between computer logic
and the design of microcomponents.

Yet in a long perspective there can be no more important and
complex instance of techniques which have both solved and begotten
problems than is provided by the history of physical communications.
Its fascination is commented on by John Snell in his introduction to
Mechanical Engineering: Railways, the companion volume to this.
Concerning its importance it is enough to point out that, of the
series of monographs of which both books form part, no fewer than
five are devoted to this most universal of 'service' industries and
virtually every other must make reference to canals and roads, tramways and ships. For from before the invention of the wheel through
to the age of space travel the need to carry raw materials, finished
products and men has done more than pose its own challenges for new
load-bearing materials, new techniques of civil and mechanical
engineering and new methods of organising labour as well as for
novel propulsive systems. It has also acted as a force shaping the
whole technical structure of a civilisation.

Through its intimate relationship to society too, transport has proved a faithful mirror of the arts as well as the crafts of the culture about it, and has even occasionally reflected back on them. Thus, nothing could be more characteristic of the easy elegance associated with the period around 1800 than a typical canal bridge, nothing express the confidence of the early years of the Victorian age better than a Stephenson viaduct. The stations of the 1840s are generally disciplined and delicately-detailed: those of the 1860s are too often heavy and extravagant. And today's blend of technical arrogance and aesthetic nonentity is epitomised in every motorway cloverleaf or airport terminal block.

It is perhaps this intrusion of cultural considerations into what is at first sight a purely technical subject which explains why the current remarkable interest in the industrial artefacts of the past should have begun with transport and remains largely centred on it. For the archaeology of industry (like almost every other branch of archaeology) commenced as an amateur movement and only later attracted the professional apparatus of lectureships and libraries. Its appeal is to the imagination—to the wise man's sense of that past on which he stands—as much as to intellectual curiosity.

And of all forms of transport, it is railways which have most captured the public attention. This attraction is not merely nostalgic, though over a century ago the term 'archaeology' was used of the tracing of a tramway; for even before 1830 the opening of a new line was regarded as an excuse for local and national celebrations accorded to few other forms of technical achievement. Ocean transport manifestly existed in its ancient and hazardous element and the canals seemed to generate their own: road improvements, on the other hand, lacked (as they still lack) the appeal of the slightly exotic. But in Britain as elsewhere—if most particularly in the country of its birth—the railway has been regarded as far more a part of the public domain than any other device with so complex a technology.

For this reason alone, Britain's railways have attracted an immense bibliography. This has admittedly been largely devoted to their most glamorous department, that of locomotion, which is not dealt with in the present book. But for more than a century historians have been so diligent in extracting even the trivia of company minutes and engineers' diaries that one is tempted to feel that there is not a yard of track left in Britain whose history is not accessible from recently published sources.

Some years back, however, the author was surprised to discover that there was no general account available of the development of the *structures* of Britain's railways—of the bridges, tunnels and earthworks which mark our landscape as characteristically and almost inobliterably as do its prehistoric remains, and of the stations and other buildings which afford so sensitive an index of contemporary taste. Since then he has himself attempted a catalogue of major railway relics. But the editor and publisher of the present series rightly felt that there was still a need for a book which should fit into its general format by providing a survey of the railway-building art—an art which was virtually synonymous with civil engineering for perhaps the three-quarters of a century following 1825, and which throws long shadows backward to Elizabethan times (and beyond) and forward to the present day.

This book is neither an exhaustive guide to the lives and works of the classic railway builders nor a full account of the expansion (and contraction) of a national network. It does not claim to mention every minor innovation, it deals with the economic background of the railway age only *en passant* (and with the politics of its warring companies hardly at all), and above all it makes no pretence at original research, reference to primary sources or informed arguments as to details of priority. But even a straightforward retelling of the building of Britain's earlier railways would be indispensable in any series claiming to deal with the background to industrial archaeology. For no comparable achievement so affected the life of a nation and of the world, and none left so many major monuments after that period of the 1850s in which a French economist first realised that something had happened in Britain in the previous half-century to which the term 'revolution'—in itself a new word—could fitly be applied.

The pages which follow, though, have been written to provide a background for those interested in studying an age in the three dimensions offered by its greater (and sometimes its lesser) structures rather than as a compact guide to early railway history. Thus the 'battle of the gauges' was far more important historically than was the 'atmospheric' railway; but the latter has left more in the way of tangible witness, and hence the two subjects receive almost equal treatment here. Similarly, it seemed better to draw attention to that wealth of minor stations which still testifies to the virtue of the last age of architecture to respect local materials than to write one more threnody for the scandalously demolished Euston arch.

Between the middle of the eighteenth century and the end of the nineteenth, Britain's tramway builders and railway builders were responsible for over 100,000 substantial structures standing in the border country between civil engineering and architecture. Many were duplicatory, but many remain unique for their technical or aesthetic significance. If this book helps towards the appreciation and hence preservation of such monuments it will have been worth while for that alone, for it is at least arguable that the postwar years have been marked equally by a passionate zeal for the preservation of minor locomotives and a reckless destruction of major railway structures. But its first purpose remains that which has been suggested above, which is to present the physical history of Britain's railways as a movement which still shapes our land and its life.

CHAPTER ONE

Prehistory

If the transport engineer is defined as a man who deals with media and routes, then his two natural enemies are gravity and friction. On the open seas the former can be disregarded and even the latter is of little importance at low speeds; and hence throughout the history of civilisation, from its roots in the Near East some six millennia ago until a few generations back, coastal navigation was, in spite of its hazards, used wherever possible for transporting heavy loads. For the inland Englishman building a neolithic temple, a Norman cathedral or a mansion in Hanoverian times, it was almost a secondary consideration whether his stone had to be shipped from Carrara, Caen or Caernarvon. The final miles over dry and hilly country were in any case likely to be the costliest.

For the same reason, early technologies exploited natural waterways to their utmost. Here gravity found an equivalent in currents; but these could be partially offset by capturing tides or controlled by weirs and gates, and to the comparatively constant levels (and hence efficient use of the limited haulage power of man and beast) which rivers afforded could be added the advantage of a smooth transit for passengers and fragile goods. It was for ease of communication as much as for climatic reasons that early centres of civilisation tended to lie in river valleys.

It is of the nature of what has come to be called technology, however, that it depends on materials which occur in localised and often inaccessible sites—metallic ores, for instance, and other specialised minerals such as coal. Accordingly, when the systematic exploitation of these began in renaissance times (and the prime incentive then, as in so many fields of technical advance, was the demands of warfare) it was shaped by transport considerations. Thus, what can be claimed as the world's first chemical-industry plant was the gunpowder factory which Queen Elizabeth I established where home-gathered nitre and imported sulphur could be carried up the river Lea to meet the charcoal of Epping Forest. Britain's ironmaking, too, tended to flourish in areas such as the Forest of Dean where there was navigable water close to its essential raw materials.

But throughout Europe there were mineral resources more or less remote from natural waterways yet too attractive to be ignored. That the exploitation of these posed such problems in the seventeenth and early eighteenth centuries was due largely to a neglect of transport engineering which had continued throughout the thousand years since the Roman empire had collapsed and grass had begun to cover its grand military highways. In a politically fragmented civilisation there has been little incentive to forge long-distance communications of any type; and now that sound roads were needed for industrial purposes there were few men available with the knowledge to survey and build them on even a parochial scale.

For their heavy inland transport, though, the engineers of the technical renaissance or early industrial revolution did not look for precedents to classical antiquity but rather made a new start. For trunk routes they still entrusted themselves to water, with the chamber-locks developed in Italy, in the Low Countries, and on such French waterways as the Languedoc canal (the first constructional work on which gunpowder was employed) pointing the way to that period between about 1760 and 1825 in Britain which is justly called the canal age. It is true that irrigation canals built in pre-Christian times often had wide sections and long tunnels, and that major navigational waterways were not without precedent in the ancient world: an ancestor of the Suez, for instance, would probably have been opened in the age of the Queen of Sheba had it not been for political considerations, and Nero began work at Corinth. Huge canal schemes were also achieved by the Chinese around the twelfth century AD. But it was the English engineers from James Brindley to Thomas Telford who (as another book in this series describes) took water across watersheds to form the most ambitious network of highways since the time of the Caesars.

The techniques which were—largely empirically—developed in this period fall into two classes, the first of which comprises methods of survey. In the later canal age the military threats of the Napoleonic wars gave an incentive to national map-making, with the Ordnance Survey beginning work in 1791: tithing and enclosure Acts, too, increased a knowledge of topographical features. But in general the engineer of 1800 had to cover virgin land with the aid of his Gunter's chain (introduced in 1624) and of the theodolite and triangulation methods which dated from the late sixteenth century, but *without* anything on paper much more reliable than a road strip-map. With such devices he had to cross not merely the gentle Trent/Mersey watershed but the Pennines themselves, looking always for the best compromise between the beeline route more suited to a marching legion than a watercourse and that devious following of a set contour which had its own balance of advantages and drawbacks in both construction and operation.

Then, its route having been chosen, the canal had to be built, with its cuttings and embankments, its tunnels and its too-frequent 'accommodation' bridges. Today computers are set to work on the type of logistical problem which Brindley solved correctly and almost instinctively. As the canal age progressed so did confidence in earth-moving, and 'cut and fill' calculations were made allowing more daring embankments and cuttings, and hence the grouping together of locks. This book is not concerned with waterway engineering as such, since that has already been dealt with by the editor of the series. But it must make so many cross-references to the vocabulary of the pioneering British canal-builders that one point should be stressed at the start.

This is, that water is a very hard taskmaster for the transport engineer. For the man on foot or horseback an extra hundred feet up and down on the traditional rolling English road (and probably rather tougher gradients on a Latin beeline route) were only an annoyance, while passengers in unsprung coaches and wagons might well be glad of the chance to get out and push after hours of jolting over rutted dirt roads. But ten feet up and ten feet down on a canal implied (quite apart from the difficulties of water supply and of capital cost) a locking delay equivalent to at least a mile added to the route. A man who had surveyed a canal hence knew how to avoid presenting gifts to gravity.

Yet though some very unpromising rivers were impounded at the height of Britain's canal age, waterways could not reach everywhere; and they were particularly ill-adapted to the needs of the mining engineer whose first need was to shift ore through sloping under-ground galleries, and whose second was probably to carry it down a hillside to the wharfs of a navigable river. Hence the mine generated another type of transport medium designed for comparatively heavy and regular traffic. Where gravity could not be defeated it was accepted; but friction could be reduced by a factor of ten as compared to haulage over rough and muddy tracks by the use of wagons with flanged wheels which engaged with well-aligned wooden rails. The latter might themselves be supported on a prepared substructure.

Another advantage of this system was that it afforded guidance. It is, indeed, probable that the first 'railways' were self-generated by the process of vehicles naturally sliding into the ruts formed by their predecessors, and may thus have been even older than the wheel. It is certainly hard to look at illustrations of the pyramid builders hauling their massive stone blocks on sleds without envisaging the discovery that the grooves of compacted sand so formed provided an easier path for the next load than would virgin land.

This is speculation; but there is firmer evidence that grooves to accommodate sleds were deliberately cut in Malta about 2000 BC and that in post-Homeric times the streets of Syracuse were paved so as to incorporate slabs for the smooth passage of wheeled vehicles. (A memory of this practice can still be seen in cobbled side-streets in the north of England, and such insets are common on the Continent and particularly in Spain.) It is less easy to be certain where the grooves in Roman pavements were consciously formed and where—as is probably the case in the forts along Hadrian's Wall—they were accidental in origin and later adapted for guidance, as were the ditch-scoured highways of pre-Macadam Britain. But there is no doubt that the rutway is far more venerable than the railway proper, which it anticipated as a guider and a reducer of friction.

There is no evidence of the use of raised rails anywhere before the sixteenth century; for though it is difficult to believe that the great Norman and Gothic churches and castles were built without some form of temporary contractors' tracks, and inclines are believed to have been used as early as the twelfth century, no guides to the building techniques of the Middle Ages survive comparable to those provided for design principles by the notebooks of medieval architects. Even the first railways which *are* hinted at appear ambiguously in words and indistinctly in woodcuts in the works of such writers as Sebastian Munster and Georg Bauer (Agricola), who illustrates a system in which wagons were guided by pins sliding in slots. All that can be said for certain is that these would be very short inter-works conveniences.

Before 1600, though, mining engineers were using wooden tracks over runs of at least a few hundred yards in three types of situation. These were underground, where the guidance of the tracks was specially useful in averting collisions: on comparatively level ground (where, as in the mines themselves, the rails were regarded as portable as with modern contractors' tracks): and on the more permanent inclines which as yet had no counterbalancing arrangements. Typical sites were the copper, tin and lead mines of upper Saxony and lower Hungary, that basin of central Europe which also saw the invention of other early mining machinery such as drainage pumps, and of blasting for ores.

Meanwhile England's mines were posing their own drainage problems, and towards the end of the century Queen Elizabeth I invited German engineers to advise on these. There is a supposition that the Germans also introduced to Britain the idea of the mineral railway. But the first of which there is certain documentary evidence appears as a home product at the beginning of the seventeenth century. Its proprietor and engineer perhaps deserves a wider renown than he has received, even though his ambitions eventually overreached his finances. His name was Huntingdon Beaumont, and even at that early date he had widespread colliery interests. Perhaps the most sophisticated of his mines was at Wollaton outside Nottingham; and there, in 1603 or 1604, he opened a two-mile line of wood-framed tracks leading down from the pit head. The place is today known as Wollaton Lane: there is no memorial, but this *is* the oldest railway site in the world to be certainly established.

Possibly as a result of his success, Beaumont extended his interests to Northumberland, and about 1608 opened a colliery system in the area of Bebside. Meanwhile, though, two other mineowners had built a line leading to the Severn at Calcutts in Shropshire, another centre of Elizabethan coal-mining. It is recorded that this innovation, which cost about £100, more than once attracted the attention of machine-breakers.

A hiatus followed, but from the 1640s onwards a number of 'wagon-ways' were constructed to serve the coalfields of the Tees, Tyne and Wear in Northumberland and County Durham. This type of construction became so typical there that it was referred to as a

'Newcastle road' and began to attract foreign sightseers. Three
centuries ago, the north-east of England was thus acquiring a reputa-
tion as the birthplace of the railway. The claim can be challenged,
though, for about 1698 an independent tradition began with Hum-
phrey Mackworth's opening of a line to serve his works near Melyn in
Glamorgan, where he had also constructed a short tidal canal. South
Wales, too, disputes with the north-east the honour of giving to these
light mineral railways the name of 'tramways' or 'tramroads',
which seems to derive from the old stem *dram* (meaning to draw)
and which was certainly used before the age of Benjamin Outram.

What the north-east can undoubtedly claim are the oldest
physical relics of the tramway age. Earthworks on Ryton Moor in
Co: Durham dating from before 1663 (and hence by far the oldest
of their kind in the world) can still plainly be recognised; and nearby
on Tanfield Moor there are two really impressive monuments built
as the coal nearest the rivers was becoming exhausted and mining
moved inland. One of these is the Beckley Burn embankment,
which is several hundred yards long, some hundred feet high, and
three hundred feet broad at its centre: dating from about 1725, it
was in use until very recently. Even more notable is the 'Causey arch'
near Tanfield Lea on the same line, which was constructed by a consortium of colliery owners with the fine title of the Grand Allies.

This incorporated a light span which immediately proved inade-
quate to its loads. Its builder, a local contractor named Ralph Wood,
was hence asked to construct a masonry successor, and did this so well
that the bridge completed in 1727 is still in good shape and is
scheduled as an ancient and historic monument. Unlike the Tanfield
Moor embankment it did not long carry traffic, and by 1800 was
being regarded as a romantic ruin. Today it is so engulfed in greenery
as to be difficult to appreciate—and even more difficult to photograph.
But in winter its lines stand out. A single, plain stone arch of almost
perfectly semicircular form, over a hundred feet in span and rising
some sixty feet above its steep valley, it is the world's first major
railway structure of masonry. And though comparatively unfamed
it is a very precious national possession.

In the first half of the eighteenth century the construction of
wooden tramways proceeded fairly fast throughout the north-
eastern counties, a typical line which was to acquire later fame
being that to Mr. Blackett's Wylam colliery: built to a 5ft gauge, this

was opened in 1748. In Lancashire and Cumberland, too, tramways were laid over levelled fields, up smoothed gradients, across bridged valleys and occasionally (though this was avoided so far as possible) through shallow cuttings or tunnels of a few dozen yards. In Scotland the first line was opened in 1722, from Tranent to Cockenzie in East Lothian, and was to acquire an unexpected use in the Jacobite rising two decades later when the defending General Cope used its embankment as a rampart for his cannon on the field of Prestonpans. And far from the scenes of early industry, some tramways were laid to carry stone for the building of the Bath of John Wood and Ralph Allen.

A mile-long incline on a gradient of 1 in 10 was built through Prior Park as part of this system in 1731, and is today used as a road. This is the oldest trace of a tramway not associated with coal or metallurgy and the oldest true rail incline in Britain, for earlier examples appear to have been little more than adjuncts to the staithes of navigable rivers. It did not, however, use the principle of counter-balance, which though patented in Britain in 1734 seems not to have been employed there before 1750.

By that mid-century year there were several dozen tramway systems operative in the British Isles, though their average length was so short that they did not total much more than a hundred miles. But the metal-working industries were now increasingly coal-based, other trades such as ceramic manufacture were also thinking in terms of tons, a rising population was becoming increasingly urbanised and so demanding food brought from a distance, and in short the industrial revolution had reached its point of take-off, with the great extra demand for transport which that change would imply. Yet comparatively little had been done to improve the nation's roads and road vehicles; and hence, even before canals proved their usefulness, there were suggestions that the short and scattered wooden tramways might one day be united into an overall system.

A further pointer to the fact that the tramways were no longer regarded as simple works conveniences came in 1758 when another colliery owner—Charles Brandling, the father of a man who was to encourage George Stephenson—opened a six-mile line from the river Aire in Leeds southwards to Middleton. For this line (which has left on the ground traces of its original state including an incline, and which after an eventful history is now in part protected by the

National Trust and operated by an amateur group) was the first to acquire a parliamentary Act. According to one school of thought this entitles it to be considered the world's first railway, though the Act only ratified existing wayleaves and the Middleton system was never open to public traffic throughout.

For though the railway age proper was comparatively close at hand its time had not yet come. The forty years before the end of the century were instead to belong to the medium which *did* first realise the idea of a national network, the canals. There are several reasons for this priority, including the one that a horse could haul much heavier loads on a waterway than it could on even the best tramway. Equally important is the fact that, after more than a century and a half, the tramways still ran on timber tracks—inflammable, infestable, and quite unsuited to the loads and abrasion to which they were subjected.

These timber rails themselves came in lengths of about 6ft, and being up to 6 inches square were almost as heavy in section as the ballasted transverse sleepers to which they were dowelled at intervals of every 2ft or less. The preferred wood was at first oak and later beech. An early improvement though, dating perhaps from before the middle of the eighteenth century, was the use of softwood rails with facings of hardwood: these facings economised in material not only because they were fairly cheaply replaceable but also because they could be cut, turned round and over, and generally re-used as convenient. They also bridged the gaps between the bearer rails.

Then, as iron became cheaper, it was a natural progression to experiment with this as a facing material. The first certain dates which can be accorded to this improvement are 1761 for wrought iron (on a small scale) in the north and 1767 for cast iron in the midlands, though there is some evidence that experiments in this field were made considerably earlier.

Even in its simplest form the timber tramway had the amazingly long life of well over three centuries, for wooden rails were being recommended for new light lines as late as the start of the nineteenth century and the last examples remained in use in Welsh quarries until the 1940s: it remains to be seen whether today's track materials will still be in use by the end of the twenty-second century. As for the plated rail, the fairly recent excavation of former lead workings at Groverake, Co: Durham, which yielded up the timber track now preserved in the York railway museum, also brought to light some iron-faced points apparently installed as late as 1850. But the faced rail had its own drawbacks—for instance, water condensing between metal and timber rusted the one and rotted the other, and the retaining nails worked loose—and was at best a temporary expedient. The situation was not improved by the fact that, although wrought iron tyres had been used on wagon wheels as early as 1726 and cast iron wheels appeared only a few years later, the norm remained the all-wooden wheel. This obviously had a short life when used on iron-clad tracks, just as a metal-shod wagon would impose heavy wear on plain timber rails.

Looking about him in 1750 (or even at the 'official' opening of the canal age a decade later), the tramway engineer had comparatively few worries in the civil engineering department; for his techniques of plane-table surveying and of spade-and-wheelbarrow construction, primitive as they were, were keeping step with the rather limited demands of an age which asked of a mineral line mainly that it should not present too many counter-gradients in a route where laden wagons were almost invariably worked downhill. He was prepared to leave dreams of steam traction and of a national network of lines to the visionary. His greatest and most immediate worry concerned (though he would not have recognised the term) his permanent way, though even in that department he was experienced in ballasting his sleepers with crushed stone or clinkers and in seeing that these were properly drained.

The main problem was to devise a rail which should be tough and hard enough to survive the regular passage of trains of laden wagons yet cheap to produce. This was a problem which was to occupy inventors throughout the remainder of the eighteenth century and at least the first three decades of the nineteenth—that period of the canal age which was also the classic tramway age. It was, indeed, not finally solved until little more than a century ago. But without rails there could be no railways.

CHAPTER TWO

The Tramway Age

The pioneers of the technological revolution of the eighteenth century—at its extremes of the pure science of a Dalton and the pure commerce of a Barclay or Lloyd, as well as in the whole terrain of entrepreneurship which lay between—were to a disproportionate degree nonconformists: in particular, it would be a pardonable exaggeration to refer to this movement as 'the Quaker revolution'. One of the first of this breed of new men had been Abraham Darby of Bristol, who in 1709 had smelted iron ore with coal (a step whose incentive had been the shipbuilders' depletion of the charcoal-producing forests of Britain) and who had hence become a founder of modern extraction metallurgy.

Darby chose as his headquarters Coalbrookdale on the Severn in Shropshire. It has already been mentioned that nearby mining villages such as Calcutts were the scene of tramway activity as early as 1605; and as other coalfields, ironworks and wharfs were opened at neighbouring sites this system was extended until, a century and a half later, its total length was nearly twenty miles. Almost accidentally, it became the first *intra*-works tramway and linked enterprises which might be commercially independent of each other, though it had no special parliamentary powers. It was typical of the early years of the tramways that this localised and detached system contained within itself a variety of different gauges, all of them narrow by north-country standards.

The Coalbrookdale Iron Works (that 'very responsible and opulent company', as a contemporary called it) had early become a supplier to as well as a user of this transport system, and before the time of its founder's grandson it was casting solid, flanged iron wheels. Abraham Darby III's own advance was to popularise the use of the cast iron *rails* which had first been made by his predecessor in 1767, just over a decade before the famous iron bridge was built nearby to carry a road across the Severn.

Rather too much may have been made of this advance in some railway histories, for the Coalbrookdale bars were not really suitable for free-standing use. Only an inch or so thick, they would have required a timber backing in heavy service, and they should perhaps be regarded as very heavy protective plates rather than true rails. The position has been further complicated by the fact that the section of track today displayed in the small museum at Coalbrookdale was (despite its labelling) manufactured at least thirty years later and is of a completely different type.

Despite some evidence that a Staffordshire line used all-metal tracks in 1777, it is indeed doubtful if a clear case exists pre-dating 1791 (when cast rails 3 inches deep and $2\frac{1}{2}$ inches wide were in service at Dowlais in South Wales), and the Coalbrookdale bars should hence be regarded primarily as a forerunner. But meanwhile new tramways continued to be built. Several lines were opened in Scotland between 1750 and 1770, for instance, an example which has left some interesting little 'tunnels' (and to this day there is no clear distinction between a short tunnel and a wide overbridge) being that which carried coal to the distilleries and other factories of Alloa, Clackmannanshire. In England, Northumberland and Durham retained their lead, while a line in Cumberland produced a weigh-house at Brampton Sands—a typical example of a facility which every important tramway needed—which is dated at 1799 and is hence perhaps the world's oldest railway building. But the tramway concept also spread south through Yorkshire into Derbyshire, Nottinghamshire and Leicestershire, and through Lancashire into Staffordshire: one or two further early examples were built in Wales too.

Even so, the amalgamated lengths of all these systems would not have exceeded much more than twice that of the Trent and Mersey canal, nor the total of their engineering works greatly have exceeded that of the Staffordshire and Worcestershire canal. It was not until the late 1780s and the 1790s, with these waterways complete, the economic importance and profitability of means for the carriage of heavy freight clear, and money for construction much more freely available, that the curve of tramway building elbowed upwards and the problem of trackwork again became acute.

All-iron rails were now being either spiked directly to timber sleepers or (after 1797) mounted by way of various forms of 'chairs', these being sometimes combined with iron sleepers. Rail breakages, however, were common; and engineers fell into the trap of thinking that they could be lessened by providing still more rigid support. Hence, well before the turn of the century, many parts of Britain witnessed the first examples of those blocks of granite, slate or grit-stone to which rails were held down by one or more spikes set in holes plugged with wood or lead. These were typically some twelve inches square and eight thick and weighed a hundredweight or more, though shapes and sizes varied widely: they were usually set square to the rails, but occasionally diamondwise at close intervals.

Such unresilient sleepers in fact increased rather than lessened breakages, and usually outlived their usefulness. But they became typical of the later tramway era and continued to be laid in the rail-way age proper even in Britain, while in Germany (as on one Cornish line) they could be seen in public use late in the nineteenth century. The railway archaeologist has reason to be grateful for this use of stone blocks, technically fallacious as it was, for their presence is a clue to the presence of a long-dismantled tramway less ambiguous than the 'dark line beneath the grass . . . short tunnel too narrow to have taken road vehicles . . . small bridge over a stream or narrow embankment across a valley . . . "Tram Inn" or "Wharf House" which cannot be related to any visible remains' which Bertram Baxter adduces. Good runs of stone sleepers can still be seen, for instance, near Silkstone, Yorkshire (1809), and Chapel Milton (1802) and Pinxton (1819) in Derbyshire, as well as in South Wales. Other examples may need a little detective work.

Although the first iron 'edge' rails had been of plain rectangular section and typically cast in lengths of three or four feet weighing as little as 30 lb/yd, local variations and purported improvements appeared well before 1800. (In 1794, for instance, William Jesssop had laid near Loughborough, Leicestershire, the first rails of a comparatively modern type with an 'I' section.) The commonest device was 'fish-bellying', or strengthening the rail by deepening it between its points of support into a scalloped profile: rails of this type were used on the line of 1815 which fed the Duke of Rutland's coal-cellars at Belvoir Castle, Leicestershire, and are reported to remain in place there. A less popular alternative was to add a horizontal stiffener, and other variants included the oval-sectioned 'Wyatt' rail and the evolution of the lap-jointed Losh and Stephenson rail in 1816. In all these cases the engaging wheel was normally of a simple flanged form; but there are some examples of a twin-flanged pulley shape being used, and a system working on this principle indeed remained operative in North Wales until a few years ago.

By this time, too, experience had been gained with the specialised forms of trackwork imposed by an increasing mileage of doubled lines, branches and shunting yards, as well as by the frequent presence of level crossings. The legend that the movable-tongued points which replaced the fixed turnout of earlier days—a device which had required that wagons be lifted manually from one track to another—was invented by a Welsh ironmaster drowsing his way through a sermon is, unfortunately, apocryphal, for crude points of the modern type had been used a century before in the days of wooden rails. What South Wales *did* popularise, thanks to its geographical isolation and perhaps its independent-minded men, was the use of a quite different type of track.

For Wales and Shropshire favoured the practice of mounting the flanges necessary to align wheel and rail on the track (normally on its inside) rather than on the wheel itself. L-shaped 'plateways' of this type were probably laid in wood in seventeenth-century mines, and John Smeaton, Britain's first great civil engineer, is credited with using one at his Eddystone lighthouse in 1756. But the idea was only translated to cast iron by John Curr of Sheffield in 1787, and a permanent iron plateway laid in the open has not been certainly dated before 1794. After that the form became characteristic of the 300 miles or more of tracks which were built in the tangled valleys of South Wales during their industrial explosion just a century after Mackworth's pioneering work. By 1803, indeed, the older Welsh edgeway systems were being converted to plateways or laid with extra rails to allow for through-running. Despite the claims made for them, these plateways—which developed fish-bellied, U-channelled and other variants—were little stronger than their rivals, and advantage was rarely taken of the fact that a plateway wagon could run over quarry floors if not ordinary roads. Their pointwork was complex, they harboured stray stones, and their only real advantages were that they offered a slightly better ratio of payload to tare and that, partly by accident and partly thanks to the foresight of their builders, a uniform gauge of 4ft 2in tended to be adopted for new plateways. But they were backed by shrewd salesmanship.

In the later 1790s many dozen miles of both forms of track were built. The extreme north of England remained faithful to its edge-rail, and was generally followed by Scotland. In South Wales, as we have seen, the plateway was used almost exclusively; but elsewhere in Britain the decision might go either way, with the plateway per-haps more popular in southern England and the edgeway in the Midlands. Thus, the numerous tramway systems built around the end of the century to feed the canals of the Welsh marches were of both types.

There have been attempts to link this rivalry with the names of William Jessop and Benjamin Outram, two of the partners who founded the Butterley ironworks in Derbyshire which was by then the focus of ironfounding. Outram certainly recommended plateways whenever consulted, and Jessop's purported preference for the edge railway has been dubiously traced to the fact that he frequently disagreed with his father, who had acted as foreman to Smeaton. Unfortunately, though some documents of great interest (such as Outram's recommendation for an ideal tramroad) have survived, the technological record is still incomplete in this important period.

For instance, it is not clear concerning the priorities in a much more important matter—the introduction of the simple, rectangular *rolled* iron sections which followed Henry Cort's invention of a specialised rolling mill in 1784. But after 1808, and particularly in the north, these began to swing the balance in favour of the wrought iron which had hitherto depended on expensive hand forging. Good rolled rails were now competitive in price with cast ones but had the advantages of a resilience and corrosion-resistance which led to a much longer life: they could also be produced in greater lengths.

So it is that the tramway scene around the year 1810—when civil engineering in Britain had at last passed the state of the art achieved by the Romans, when anonymous works were becoming rare, and when an increasing number of voices could be heard in favour of standardising specifications in the interest of nationwide communi-cation—was more lively than it was coherent. Timber tracks were still being laid, and for heavy duty the choice was between the cast or rolled edgeway and the cast plateway, with the rolled plateway soon to follow. These sections might themselves be supported on various combinations of wooden and cast-iron sleepers and stone blocks, might be laid to a dozen or more gauges selected on the basis of little more than local whim, and might differ in other ways which made through-working impossible. Figure 1 illustrates a few typical rail sections, but at least fifty variants were used in Shropshire alone.

That despite such drawbacks the tramway system continued to expand can be traced to the accelerating demands of the industrial revolution. That it continued to expand comparatively smoothly is due to the fact, recognised by Telford in 1800, that tramways and waterways were to a large extent natural complements of each other. The canal was attuned to heavy tonnages, fairly long hauls and level country: the tramway was more suited to the lighter load, the journey of less than ten miles and (with the aid of a plain or counterbalanced incline) the hilly terrain. The media frequently shared engineers and owners, it was rare for a waterway to be converted to a tramway or *vice versa*, disputes as to which should be constructed in a given situation were almost unknown, and certainly the two never ran in such competition as was later to develop between canal and railway companies. Indeed, early parliamentary Acts often simply approved a route and left open the question of whether a tramway or a canal was eventually to be constructed.

Hence many of the tramways built around the turn of the century (and, indeed, as early as the 1780s) were feeders to the waterway system, a fact which helped their operators in that loads were still normally carried downhill. A good instance of this type of line, and one of which remains are not hard to find since it operated well into the present century, is Outram's own Peak Forest plateway. Opened in 1797 from the terminal basin at Bugsworth, Derbyshire (now politely renamed Buxworth), to limestone quarries near Dove Holes, this operated on a twenty-four hour basis. It is also of interest as an early example of integrated transport: the same company owned quarries, canals and tramways, and iron containers were used which could be easily transported from wagon to boat. This seven-mile line had a self-acting incline near Chapel-en-le-Frith and a hundred-yard tunnel—the longest even partially surviving from the tramways of the eighteenth century, and possibly the first true railway tunnel ever built—at Chapel Milton.

Another typical and long-lived tramway which fed the Ashby-de-la-Zouch, Leicestershire, canal (1802) is noteworthy for what is

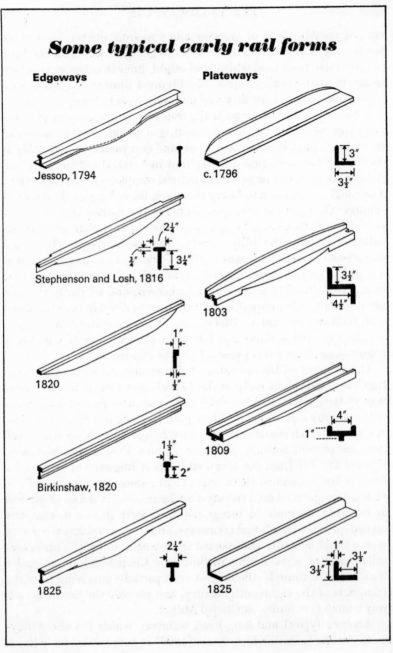

Some typical early rail forms

Edgeways

Jessop, 1794

Stephenson and Losh, 1816 — $2\frac{1}{4}''$, $\frac{3}{4}''$, $3\frac{1}{4}''$

1820 — $1''$, $\frac{1}{2}''$

Birkinshaw, 1820 — $1\frac{1}{2}''$, $2''$

1825 — $2\frac{1}{4}''$

Plateways

c. 1796 — $3''$, $3\frac{1}{2}''$

1803 — $3\frac{1}{2}''$, $4\frac{1}{2}''$

1809 — $4''$, $1''$

1825 — $1''$, $3\frac{1}{2}''$, $3\frac{1}{2}''$

Figure 1

perhaps the first of 'skew' bridges, and another interesting early bridge (1795) can be seen near Little Eaton, Derbyshire. Though the level central Midlands preferred to extend their canals directly whenever possible, there was soon hardly an industrial waterway from the Leeds and Liverpool in the north to the Somerset coal canal in the south which did not have its cluster of associated tramways.

Many navigable rivers acquired feeders too, a good example of this type being the devious, 26-mile Plymouth and Dartmoor edgeway of 1823. Yet other tramroads led directly to deep-water harbours such as Seaham and Whitehaven in the north and Poole and Devoran in the west. In this group a Scottish plateway, the Kilmarnock and Troon of 1811, has left two impressive stone bridges, that spanning the river Irvine having five 40ft arches. The line can also boast of being Scotland's first public railway, and like England's first (opened eight years earlier) was engineered by William Jessop.

The nineteenth century also saw the inception of a new variety of system in North Wales: the edgeway, usually laid to a gauge of under 2ft, serving slate quarries. The first such began at Port Penrhyn (1801) and the most famous was to develop from a contractors' line at Portmadoc (1811). But, as has been mentioned, it was South Wales which witnessed the most intense growth of tramways in the period 1790–1820, frequent advantage being taken of the 'eight mile law' which gave a canal the right to operate lines that far from its banks. Indeed it was not long before there was hardly a valley in Monmouthshire and Glamorganshire which did not have a plateway as well as a canal to link its pits and quarries, by way of its forges, foundries and mills, to its cities and harbours. Representative lines can also be traced in Carmarthenshire and even Pembrokeshire.

Several of these systems (which were often interlinked both physically and by their toll systems) have left particularly interesting remains, such as the fine and almost identical semi-elliptical stone bridges of 1815 at Edwardsville and Greenfield on the historic, and older, Penydaren plateway which are each more than 60ft long, the five-arched span near Port Talbot and the later three-arched ones at Glan Rhyd and Carnavon, Glamorganshire, and the little iron bridge of 1811 near Robertstown, Aberdare, which is believed to be the world's first free-standing rail bridge of metal. Inclines were numerous, some ingenious machinery was devised for loading and trans-shipment, at least one 'floating bridge' was built, and another curiosity which shows the tramway engineers facing an unusual problem is the 'sand tunnel' which can still be traced on Porthcawl beach.

Near Blaenavon a tunnel over a mile long survives from an otherwise unremarkable line laid in 1815. Typically small in bore (and brief by canal standards) as this is, it is easily the longest tunnel of the tramway age if one discounts those which were essentially extended mine galleries. The great viaduct at Risca, though, has vanished save for one abutment.

Other Welsh plateways penetrated north into the Black Mountains, the Brynoer line of 1815 showing a particularly fine example of contour-following engineering; and in 1817 these lines were extended into England by way of the 24-mile Hay plateway which ran from the head of the Abergavenny canal at Brecon to Eardisley in Herefordshire and which has left some major relics including a tunnel of nearly 700 yards. A similar complex (whose most striking monument is the iron 'Waterloo' bridge of 1820 over the river Arrow) extended from Govilon, Monmouthshire, to Hereford, and yet another which has left relics operated some dozen miles of track in Monmouthshire.

Furthermore, it is not unreasonable to regard certain plateways beyond not only the Usk but the Wye (and even Severn) as belonging to the tradition of South Wales. The Forest of Dean, for instance, developed a complex network of public lines of its own: this included a tunnel of over 1,000 yards at The Haie which was later, like much of the mileage in Wales and the west, converted to conventional railway use.

Another function of the tramways was to act as temporary 'bridges' during the construction of long canal tunnels or lock flights. (This was, of course, distinct from the building of the various forms of barge lifts and inclines which have left their own traces on the land.) Both the Blisworth tunnel on the present Grand Union canal and the Marple lock flight on the Peak Forest canal, for example, were preceded by steep tramways which operated for a few years whilst more permanent works went ahead: a tramway was also built at the Devizes locks: and the sharply locked Northampton canal arm was originally entirely a tram road. Much longer-lived was the line which joined the two never-connected sections of the Lancaster canal across the valley of the Ribble at Preston and over Walton

Summit, which is believed to be the site of the world's first power-operated incline.

Despite such spreading tentacles the transport industry of the early nineteenth century remained a closely interknit affair. Miscellaneous trades, including fairly 'light' ones such as glass-making, and agriculture too, made their contribution to the traffic of the tramways and canals, while other commodities which might appear on waybills ranged from moulding sand and china clay to pit props and zinc ore. But all was largely based on coal and iron, and even within that world there was a high degree of cross-fertilisation so that a company which was a major *user* of the tramways was probably also a *supplier* to them as well as a manufacturer of steam boilers and of plates for canal aqueducts.

The earliest years of the nineteenth century, though, had seen another step towards the realisation of the modern concept of the general-purpose railway with the opening in 1803 of the Surrey Iron railway and its immediate extension, the Croydon, Merstham and Godstone line. The first tramway of any length in the London area, this sixteen-mile, double-tracked plateway laid to the 'Welsh' gauge of 4ft 2in had been planned as a canal and was in essence a waterway feeder, carrying the limestone of the North Downs to the Thames. But it was neither formally nor in practice restricted to such traffic; and the historic importance of this major contribution by the south to the tramway scene is that the Surrey Iron railway, like the canals themselves, was given by parliament the rights and responsibilities of a 'common carrier'. It was not an adjunct to a manufacturing concern but a transport enterprise in its own right. And as such it is another claimant to the much-disputed 'first railway' title.

The engineer of this 'vast and important concern' was William Jessop, who built (in addition to other early railways mentioned here) a number in Leicestershire and who was later assisted by his son. Its contractor, Charles Joliffe, is associated with such other important works of the period as Rennie's London Bridge. Good remains of the southern extension still survive, and it is said that the rockeries of Wandsworth are particularly rich in stone sleeper-blocks.

In this opening decade of the nineteenth century the demographic effects of the innovations of the previous twenty-five years were beginning to be felt in terms of a demand for reasonably fast and smooth passenger travel as, to an increasing degree, a man's place of work ceased to be synonymous with his home. Trunk routes such as the Bath, Dover, Holyhead and Edinburgh roads were at last realigned and macadamised, and with that frenetic work at the changes of teams which characterised the classic stage-coach era 15 mph averages were attained and Brighton brought within daily commuting distance of London—at least for stockbrokers. On the canals a few 'fly' passenger boats achieved speeds of up to 10 mph, and much lower fares. But lock-free canals were hard to find, and even roads permitting the best to be got out of a horse were so rare that there was a constant temptation to country hauliers to work their wagons along the roadbeds of the tramways.

After more than two centuries of tramway operation the only passengers other than opening-day officials to have travelled over rails (and those at best semi-officially, as in coal wagons returning empty) were works employees. When at last a line acquired powers to transport the general public it indeed seems to have done so almost absentmindedly. This line—another with a 'first railway' claim, and certainly the first fully public one in Wales—was a seven mile plateway, with the pleasant name of the Oystermouth railway, which crossed the Mumbles peninsula from Swansea. It was opened in 1806, started a regular passenger service in the following year, and over the next century and a half enjoyed so varied a history that it is tragic that it has now been almost totally effaced.

Passenger haulage reached Scotland five years later; but its spread even in England was very slow and not until the late 1820s did it become a factor in the planning of a transport system rather than an afterthought for an existing line. (The growth of *public* tramways, on the other hand, was comparatively rapid, with a new Act being passed on the average every year throughout the 1810s.) So, as Britain began to emerge from the post-Napoleonic depression, it was to find itself still in the canal and tramway era, with the only conspicuous technical innovation being the Durham engineer John Birkin(or, incorrectly, Berk*en*-)shaw's invention in 1820 of a process for rolling I-form iron rails about 1in wide by 3in deep

These were soon afterward available with a fish-bellied base. They were produced in lengths of up to 20ft and eventually reached a weight of 80 lb/yd, with even continuous welding proposed now that the shock-creating effect of short rails was fully appreciated; and their appearance marked the beginning of the end for new plateways. For of the fifteen lines proposed in 1825 and 1826 only two (in South Wales, of course) were of this type.

That Birkinshaw rails so soon became popular for the richer companies was largely because of the fairly successful experiments in steam traction which had begun in 1804 on the Penydaren plateway and had continued on the edgeways of Tyneside. These experiments are dealt with more fully in the companion volume to this book. But they had brought to a head the problem of the brittleness of cast iron rails, whose frequent fractures had limited the weight (and hence adhesive power) of early locomotives hauling unsprung stock and had thus severely restricted the development of the steam engine itself.

This account has now reached the period when civil or non-military engineering was emerging as a recognised profession, with an institution founded in 1818 giving it a respectability which the rude mechanicals (who were at first regarded as little more than unqualified hobbyists) were not to attain for another three decades. Other changes were in the air around 1825; and so, with the tramways now totalling close to their peak length of some 1,500 miles and operating in the majority of English and Welsh counties, it is worth taking a general backward glance over a scene then character-istic of Britain.

Had this book been published fifty or even twenty-five years ago the material contained in the last score of pages would probably have been compressed into two or three; for it was generally held by not only Victorian historians but their successors that the railway age began suddenly in 1830 (or perhaps 1825) and that before that date only the pioneering of steam locomotion was very relevant to later development. But if recent years have seen a great increase of in-terest in railway prehistory, this is only in part because the visual aspects of the early industrial revolution—the proud, five-storey waterside warehouses and the two-man forge-hammers hidden in lonely woods, the serene canals and the tramways with their horses patiently plodding throughout a long day to haul a single wagon the width of a county—appear so attractive in our hasty times. It is also because the concept 'railway' is now seen to depend as much on civil as on mechanical engineering, and because the former came of age between 1760 and 1825.

Between any canal built by Brindley at the start of this period and any built by Telford towards its end there is an obvious advance in boldness of conception and (generally) in assurance of execution: the main technical improvement was, of course, Telford's own introduction of 'cut and fill' calculations. The tramways as a whole do not exhibit this progress; for they were far more bound to local needs and resources than the waterways, and many permanent lines opened late in their age were still laid on ground only roughly levelled. But *at its best* the new or realigned tramway of the early nineteenth century was as skilfully engineered as the contemporary canal, with its gradients restricted to between 1 in 50 and 1 in 100 to allow the haulage of trains of up to fifteen wagons (or 50 tons) by a single horse, its road-bed 12ft wide for a single line and 18ft wide for a double one, its embankments up to half a mile in length and 50ft or more high, and its attention to drainage. Only tunnels were still treated as a major difficulty, with Outram confessing that 'in very rugged countries short tunnels may sometimes be necessary' and bores being restricted to as little as 7ft. It is arguable that it was only a lack of the capital and the challenge which prevented the leading tramway engineers such as Outram, the Jessops and the Rennies, and such less-renowned men as John Hodgkinson, John Wilkinson and George Overton too, from leaving memorials on the landscape to compare with those of a Stephenson or Brunel.

But as it was the railway age had long remained poised at its point of take-off, thanks in part to economic factors and in part to the persistent, interlinked problems of developing a workable locomotive/rail system. Thus as late as 1802 an extraordinary line was laid from Haytor in Devonshire whose 'rails' and even 'points' (many of which still remain in place) consisted of rabbeted granite slabs, while other oddities of the period included two patterns of monorail. And so the tramway age appears as more divorced than it was from the railway age proper, and it is often forgotten that it has bequeathed to British technology such familiar terms as 'sleeper' and 'platelayer'.

This scene of a century and a half ago is not, however, accessible only to specialised students. Until recent years the actual metals of both early plateways and edgeways remained scattered fairly representatively on the face of Britain and could even be found in works usage, but so much has been or is in danger of being lost that a list of *in situ* survivors might lead to disappointment. In come cases lengths have been preserved and re-erected in sites near their original ones, and in others they have been rescued from such tasks as serving as cattle barriers; but even the museum preservation of rail sections does not present a completely satisfactory picture, and these un-glamorous lengths of scrap-iron are too often mislabelled before being relegated to euphemistic 'reserve collections' in obscure basements.

A good many examples, too, remain in private hands. The situa-tion is all the more frustrating in that rail sections tend to survive in considerable lengths, so that there is usually enough to provide representative specimens for a private enthusiast, a local museum and a central collection too. Whilst ambitious schemes for national transport museums are being argued out, a definitive group of rail sections could be assembled for a few thousand pounds. Meanwhile, one of the best collections is reported to be still lost after a trade fair in Cairo half a century ago.

As has been mentioned, though, a reasonable number of runs of stone sleeper blocks survive (occasionally with their spikes still in place), and even such oddities as mile-posts can still be found *in situ*. There are more than a dozen tunnels exceeding fifty yards sur-viving from before 1825 (though several up to a thousand yards in length have vanished utterly), a few cuttings, and a larger number of embankments. Perhaps a score of major underbridges or parts there-of still stand, as well as many minor bridges and culverts. Inclines are easy to recognise if hard to date. And above all the tramway age subsists in its routes, which still challenge to new discoveries the amateur willing to blend library and field research—and which still come to light accidentally too.

Some of the tramway mileage was converted into conventional mineral lines or even general purpose railways. Much more has been built over, particularly in recent years; for man has done more than nature to remove the evidence of an age. But several hundreds of miles of untouched tramway roadbed remain, often in country deserted by modern industry. Whether through the highlands of the Pennines or Beacons or in the green Kiplingesque shade of lost ways through the woods, these often provide walks splendid for their own sake. But, like the Roman highways and prehistoric trackways which they so closely resemble (for 'bronze age castle', read 'embankment': for 'villa', read 'wharf house'), they are walks through more than distance and lift a topsoil off time to reveal the ways of a vanished Britain.

CHAPTER THREE

Years of Transition

The distinction between the typical tramway and the typical railway is only in part one of materials and techniques. Equally important is the question of *scale*, of whether a line is considered as a linking of facilities on an almost parochial level or as part of a national system.

This is also (since transport is so sensitive an indicator of cultures) one of the fundamental differences between the medieval and the modern concept of Britain. It would be wrong to equate the coming of unity with the development of railways, however, for the canals took this step in the few years between the completion of the Bridgewater and of the Trent and Mersey canals. In its brief heyday between 1820 and 1840 the stage coach too was, with all its limitations and drawbacks, helping to weld the nation together. And in fact the 'through route' concept was remarkably slow in seriously affecting railway building.

As long ago as the planning of the Surrey Iron railway, William Jessop had proposed that this line (which in fact never reached its intended terminus at Godstone) should be prolonged to Portsmouth, a place of great importance in view of the threats of French invasion and channel blockades. A survey was indeed carried out over the full distance from London to the coast. But all that eventuated was a rather unsuccessful canal, and little more was heard of the idea that the tramways could act as other than works conveniences or feeders to waterways until 1819.

Then, however, a vigorous campaign on behalf of the railway as an inter-city artery was begun by William James, who had conceived the idea as far back as 1808. A publicist rather than an engineer (for instance, he became more successful in popularising Birkinshaw's rails than in selling his own hollow variant on them), and perhaps a Victorian born twenty years too soon, James surveyed at his own expense the length of a line which should link, if rather indirectly, London and Birmingham. All that came of this was essentially another horse tramway, built in 1826 as an edgeway using Birkinshaw wrought iron rails laid in chairs on stone blocks to a gauge of 4ft 8½in; but it is a good example of its type.

A canal in which as a local man James held interests existed south from Birmingham to Stratford-on-Avon. Taking over from there (where the typical weigh-house survives), James built a bridge which must be regarded as the most elegant of all survivors of the tramway age and one of the last things to be added to the face of Britain which show man and nature in complete harmony. Its nine semi-elliptical brick arches, each of 35 feet span, still firmly and gracefully cross the Avon and carry far more pedestrians than recognise their origin. The designer of this fine thing—born in Morpeth and buried in Brighton, but then resident nearby in Stourbridge—was John Urpeth Rastrick, a talented all-round engineer who is credited with inventing the sight-glass water gauge for boilers.

South of the river the route has left earthworks beside the Alderminster road and a number of bridges, cuttings and minor embankments on its sixteen-mile course through the Red Horse Vale to Moreton-in-the-Marsh. There it expired, having done little more than provide an outlet for the stone and agricultural produce of the area. Later the southern section of the line (and a spur to Shipston-on-Stour) was converted to a railway branch; but the northern section was abandoned, and Stratford never became a stop on the way to Birmingham.

It is clear, though, that before 1820 Birmingham had marked itself out as one of the great manufacturing centres of Britain, with a population of 100,000: it had few natural advantages, but to the Black Country's position as the crossroads of the waterway system had been added the enterprise of a group of outstanding engineer-capitalists such as James Watt. As the second city of the industrial revolution, however, Birmingham was challenged by the older Manchester. And whereas the canals provided a reasonably direct route from London to Birmingham, Manchester's connections were poor in a direction south-east of the Pennines. As late as the 1820s schemes were proposed for a water link between the Peak Forest and the Cromford canals, but this would have had to follow a very indirect route. For between rose the massif of the High Peak, over

1,200ft above the sea and 750ft above canal levels, a desert of water-less, fissured limestone.

In the event, therefore, a tramway was built over the thirty-four miles between Whaley Bridge and Middleton on the two canals. The engineer was Josias Jessop; but despite the date of 1830 and the use of the 4ft 8½in gauge which was now becoming standard, short and fish-bellied cast rails supported on stone blocks were still used. The only local produce for the line was limestone, and it should rather be regarded as another 'bridging' tramway—the last of its kind, and perhaps the most remarkable. Its cost of £150,000 was only a quarter that of the proposed canal.

Even its course as it zigzags across the heights between drystone walls is well worth following today for its atmosphere and its example of early engineering in difficult country: there is one gradient, for instance, of 1 in 14. There are also a number of earthworks and bridges, and two notable tunnels at Barbage (640 yards) and New-haven, Derbyshire: since the latter was bored for a double track but only used for a single one it has kept its original form, with the company's arms still inscribed on the portals. But undoubtedly the most striking features of the system were its inclines, and particu-larly the pair at the eastern terminus.

It is a mistake to associate such cableways only with early tram-way construction. On the contrary, routes with a continuous steep gradient were often realigned in the earlier nineteenth century into 'staircase' form; and (particularly when the lie of the land favoured them) the use of inclines continued into the age of railways proper on such lines as Stephenson's Whitby and Pickering of 1836, which also had the peculiarity that it was both empowered and forbidden to use locomotives. Such cableways were, indeed, made more popular by the perfection of the stationary steam engine after 1800; and the two great runs near Cromford, each three times the size of the next largest of the age, show incline building on a scale not exceeded until the cliff funiculars of the end of the century.

Each is some three-quarters of a mile long and climbs 400ft at about 1 in 8. The lower, called Sheep Pasture, is an amalgamation of two earlier inclines: the upper, at Middleton, is in its pristine form. Both preserve their polygonal engine-houses, and in the higher (which is scheduled for preservation) there survives the original winding engine. Some office buildings also subsist.

Part of the Cromford and High Peak line remained open until a few years ago, its last task being to supply water to the isolated uplands, and it is sad that no society has been formed to preserve such unique features as the startling signalling arrangements at Sheep Pasture. It was not the last important tramway to be opened, however, though the only public horse-hauled line built after 1830 (which was outside Edinburgh) became a joke almost as soon as it was completed.

In 1832, for instance, there was completed near Keynsham, Gloucestershire, a 4ft 8½in edgeway which has left some fine remains, including a rock cutting and tunnel at Willsbridge (now used to carry a water-main) and a weigh-house beside the Avon. The plateway was now becoming extinct, but other edgeways continued to be built—mainly to the standard gauge—in Leicestershire, Cheshire, Wales and elsewhere for a further two decades: one which has left an unusual relic served the china-clay workings inland of Par in Cornwall by way of the grand and unique double-decked Treffry viaduct of 1847, whose ten arches still support a working aqueduct. To this period too belongs the extension of the remarkable Festiniog narrow-gauge line in Merioneth (1836) with its continuous gradient over twelve miles. Probably the last line of all to be laid on stone blocks was opened in the Valle Crucis near Llangollen as late as 1852.

But the Stratford and the High Peak lines, taken in conjunction, have an unusual historical importance. They show that, as late as the close of the 1820s, canals and tramways were regarded as pieces in the same jigsaw of transport, and that in the latter field the cast iron rail was not yet outmoded. They show too a weakness in the lines of communication between Britain's great cities which was to endure until the end of the next decade and which, despite all road improvements, affected men as well as materials. For although the Stratford line did not regularly carry passengers, the High Peak railway (for all the delays caused by its inclines and transhipments) operated a cross-Pennine service for some years after 1834, having in the previous year acquired its first locomotive.

Meanwhile, though, yet another line with 'first railway' claims had been opened. In 1821 an affluent Quaker named Edward Pease at last received parliamentary sanction for a project which had been in the air for over half a century and had, in fact, been surveyed by George Overton. This was a common-carrier route whose main task should be to connect the collieries around Bishop Auckland, Co: Durham, to the port on the Tees at Stockton. George Stephenson was now commissioned as chief engineer, to be assisted by his eighteen-year-old son Robert.

The older Stephenson was then in his mid-forties. As a self-educated man he had been a late developer: furthermore, his first renown had been won as a mechanical rather than a civil engineer. Soon after the start of the century, certainly, he had been associated with the building of the Killingworth colliery line; but not until 1819 had he designed a railway of his own, and then it was merely another local tramway (mainly cable-worked) at Hetton colliery. But he was the man the times needed, for his experience in the Northumbrian coalfields had given him what few other leading figures of the age possessed—an understanding of, and belief in, the powers of steam. He converted William James, for instance, to the view that a nation-embracing network of iron rails could not come about without steam as its partner.

George Stephenson's trust in the adhesive (rather than tractive) power of the steam locomotive was, however, a limited one. For gradients as gentle as 1 in 100 he would accept a powered incline. And even on the level he still considered Benjamin Thompson's system of cable haulage (as used on the Brunton and Shields line of 1826, where some engine-houses may survive) as a viable alternative to the 'travelling engine'.

Hence, as opened in September 1825, the so-called Stockton and Darlington railway was somewhat of a hodge-podge: even its historic name is a misnomer, for Darlington fell short even of the half-way point on a system of more than thirty miles which terminated at Witton Park near the Wear. For the greater part 28 lb/yd Birkinshaw rails, to which Stephenson had been converted despite his commercial interest in the cast Losh rail, were used, being laid to a gauge of 4ft 8in to which the extra $\frac{1}{2}$inch was not added before 1840; but these were still bedded on stone (or, near the port, wooden) blocks, and a proportion of cast iron rails was used to placate the conservatives. Locomotives were employed for freight haulage; but the only novelty was that they were used regularly rather than experimentally, and passengers continued to travel behind horses. The line was fully 'public'; but in fact it drew almost all its trade from the Co: Durham collieries.

Even in its civil engineering works the Stockton and Darlington shows a transitional character. The section between its eponymous towns has been continually rebuilt, but has preserved an unusual three-arched bridge by Ignatius Bonomi, the first professional architect to work for a railway, across the Skerne at North Road, Darlington: of this, the south side exhibits the original work. Close by, where the line crossed the river Gaunless, there were two other interesting bridges. One (serving a branch and recently demolished) was built on the skew and gave its designers so much thought that a full-sized model had first to be built in timber to supply the masons with patterns: this device had been resorted to on unusual canal bridges, and was to be needed for later railway ones too. The other bridge was of cast iron, and despite its diminutive size was of an elaborate and even sophisticated 'bowstring' design which dimly foreshadowed Brunel's work. When it was dismantled in 1901 (leaving the abutments *in situ*) a part was mercifully preserved, and now forms one of the more interesting exhibits in the railway museum at York.

When one adds that the Stockton and Darlington also boasted the first building from which passenger tickets were sold and (at Shildon) the first railway-owned locomotive works too, it is clear that it justifies its historic fame. But on its steep and disused upper reaches at Etherley and Brusselton there can be found, together with other interesting relics, the blocks of cable inclines which might have been built two decades or more earlier, and on a branch at Stanley a winding-house survives. Thanks to Stephenson's conservatism as both a mechanical and a civil engineer, the Stockton and Darlington was at least as far removed from the lines which were to begin operating just over a decade later as these were from the modern railway.

Furthermore, despite its commercial success, the Stockton and Darlington appeared to have no immediate emulators; for even the nearby Clarence railway hung fire. At this period about half a dozen parliamentary Acts were being passed every year for new lines, some of these being essentially heavy-duty tramways of the High Peak type and others, with their provisions for locomotive haulage and passenger traffic, more truly proto-railways. But the next opening of major importance did not come until May 1830.

This was that of the six-mile Canterbury and Whitstable railway, all that materialised of William James's scheme to bypass the Foreland with a line from Gravesend to Shoreham. Despite its timber sleepers it marked no great technical advance, for though it had one Stephenson locomotive it relied more on horse traction and two long inclines. One of these passed through the 1,000-yard Tyler Hill bore, which can claim to be the first tunnel of the railway age and which remained in use until this historic line became one of the first casualties of the postwar closure mania. But the C & W did introduce true railways to the south, and was the first line anywhere to carry passengers regularly, if briefly, behind a locomotive.

Meanwhile a much more important route was approaching completion in the north. Liverpool, then as now, was Britain's first cargo port for communications with the New World, and Manchester had consolidated its position as a great centre of manufacture and population. It looked to the Mersey for its raw cotton, some of its coal, and its imported grain and livestock: back to the port it sent cloth and other finished goods. Yet the only general-purpose carrier operating between these two cities, some thirty miles apart and with a combined population of over 350,000, was still the Bridgewater canal.

Perhaps the most famous remark in transport history had been passed by that waterway's ducal sponsor when he predicted that the sole threat to its prosperity would come from 'those damned tramroads'. But its proprietors had overplayed their monopolistic hand, and in the early 1820s the tolls on the Bridgewater canal gave rise to bitter complaints by the merchants of Manchester and, even more, of Liverpool. This period marks the beginning of the end of more than half a century of harmony between land and water for inland transport, and with it the beginning of the end for the canals themselves. The child (or, at least, smaller brother) was about to swallow its senior.

Around 1800, William Jessop and Benjamin Outram had prospected the country between Liverpool and Manchester. But it was the energetic William James who revived the project of building a railway when, in 1822, he surveyed the route with a party which included the young Robert Stephenson. Soon afterwards James was in financial difficulties and Robert Stephenson overseas; and so, after his work beyond the Pennines was completed, the latter's father was commissioned to work on the Liverpool and Manchester railway. He was, however, at first only a *primus* (if that) *intra pares*, the *pares* including the Rennies, who had recommended a 5ft 6in line, and a gentlemanly ex-army officer named Charles Vignoles.

Five years later, in September 1830, there came the historic opening day. The bands and flags and cannon, the Duke of Wellington (no railway lover, he) and the sullen post-Peterloo Manchester mob, the delicious Fanny Kemble lost in admiration and William Huskisson waiting in the wings to die—these chase across the scene like the sun and storm clouds which marked the day. Bewitched by such a colourful parade, it is easy to lose sight of what was, and what was not, important about the Liverpool and Manchester railway.

Its first significance, of course, is that it relied wholly on locomotives. That it did so was almost an accident, for though Stephenson had rejected suggestions that the entire line should be worked by cable traction he had laid most of its length to an extremely easy gradient of about 1 in 1000 and had incorporated three sections which he intended to treat as inclines. The canal practice of alternating levels and lock flights still dominated the minds of transport engineers, and it was only after the Rainhill trials that it was realised that the two planes at around 1 in 100 were well within the capabilities of *Rocket*-type engines, leaving only the Liverpool terminal section to be operated with cables. But the locomotive *did* prove itself on this line; and though horse rail traction was to endure in special circumstances elsewhere until the outbreak of the second world war, the year 1830 marked the beginning of the end for this noble beast too.

Another quality of the Liverpool and Manchester railway was a certain note of professionalism in its organisation and operation— the same type of professionalism as was manifesting itself in the early works on railway engineering written by Stephenson's colleague Nicholas Wood and on transport topography by Joseph Priestley. One example of this quality is that the line boasted at its termini the first proper stations: another is the fact that only L & M stock was allowed to travel over its metals, whereas the tramways had been open to private hauliers and even on the Stockton and Darlington passengers had travelled in independently-operated coaches. The toll-gated tram roads had indeed been run almost like turnpikes; and on some of them, as on the canals, the proprietors were, through fear of monopoly, formally prohibited from acting as carriers.

An interesting speculation is on how long a national railway system could have been worked on the apparently reasonable principle of separating the functions of track-owner and carrier. In the event the Liverpool and Manchester chose (and was permitted) to fuse these completely. And in so doing it set a pattern for all the railways of the world, except in that the private *ownership*, as opposed to *operation*, of goods wagons remains a useful device even today.

George Stephenson had demonstrated his usual conservatism in taking short fish-bellied rails, weighing only 35 lb/yd but of the wrought iron which was now standard, as his normal track. These were supported on stone blocks, but an exception was made where at its Manchester end the line crossed Chat Moss. This four-mile-wide swamp, on which a man could walk only on snowshoes, remains even now an open waste amid the closely built-up conurbation north of the Mersey; and it presented Stephenson with a unique problem. He overcame this by patience as well as ingenuity, first cutting drains and then either establishing a 'floating' roadbed or (at the eastern extremity) sinking rafts of wattle and ling for month after month until a bottom was touched 25ft or more down and a stable embankment rose five feet above the surface of the bog.

To reduce the static load, Stephenson used wooden sleepers on this section as on other short runs of the line. These cross-ties, whose earlier adoption had been hindered by the fact that unless buried they were damaged by the hooves of haulage horses, proved more satisfactory than the stone blocks, which were thrust apart by the force of locomotives running at speeds of up to 50 mph unless they were expensively held in by tie-rods. So in 1837 (and after several patchings-up) the Liverpool and Manchester was relaid with 50 lb rails supported on timber sleepers throughout—except, presumably, for some sidings, as a few stone blocks are still preserved at the Manchester terminus. The gauge was 4ft 8½in, and the tracks were doubled throughout as on all important lines thereafter.

With one exception the other civil engineering works on the line were not individually epoch-making, but taken together with the Chat Moss crossing they were of a scale and a boldness never before compressed into a transport route only thirty-one miles in length. An interesting skew bridge near Rainhill and a viaduct at Newton-le-Willows still carry traffic, for example; and at the Liverpool end the two-mile Olive Mount rock cutting and 2,250-yard Edge Hill tunnel through very mixed strata (which used cable haulage and did not at first accommodate passengers) are substantially Stephenson's

TRAMWAY ENGINEERING

1 *Top* The 'Causey Arch'
2 *Bottom left* Bishopwearmouth, Co. Durham
3 *Bottom right* Belmont Bank Head, Co. Durham

THE INCLINE SYSTEM 4, 5 Hetton Colliery railway

originals, even if the former 'awful chasm' has lost its dramatic
quality through the successive widenings which led to a Georgian
church being moved *in toto* three times over. Less familiar to the
ordinary traveller is the spur which leads through steep tunnels
down to the Liverpool docks at Wapping.

But perhaps the greatest memorial of all to George Stephenson is
to be found near Earlestown. Here a nine-arched viaduct—in the
strict sense of a series of arches, rather than the more generous usage
which includes embankments—towers 70 feet above the valley of
the Sankey Brook and an early canal. This is not merely the first
such great work of the railway age proper but a prototype for a
form of structure which was to be repeated a thousand times over
before the end of the century, which was not to be finally outmoded
by new designs and methods until after the first world war, and
which remains as characteristic of the English landscape as do her
church spires and last hedgerows. It was to be varied by local materials
(Sankey itself is slightly unusual in that its brick is faced with blocks
of masonry weighing up to two tons), by the individual approach of a
Brunel, and by such long-lived regional traditions as those of East
Anglia. But in the great majority of the situations where a British
railway crosses a river valley it is on a Stephensonian type of viaduct.

This may have been designed by George, by his son, or by one of
the north-country engineers who followed his lead even after they
struck out on their own. It consists of a series of piers rising at centres of about 40ft and each some 20ft square at the base. As they
mount they taper gently until they reach their entablatures. And
from these there spring arches semicircular in elevation or following a slightly flattened semi-ellipse, with a low and probably plain
parapet to the track deck.

To appreciate just how firmly (and in spite of all his faults, for he
was neither a cooperative nor an imaginative man) George Stephenson stamped this sign on the face of Britain one must—as in general to
see one's homeland in perspective—travel overseas. Only a year after
the opening of the Liverpoool and Manchester, America's first public
and passenger-carrying line was operative, instituting a tradition
largely independent of the British one. In the later 1830s the railway builders were to be at work on the continent of Europe, and
there, too, styles attuned to local origins and materials were soon
to develop in several different countries. The masonry bridges on

Austria's Semmering line of 1844, for instance, appear to owe almost nothing to the tradition of Causey and Darlington and Earlestown (though metal structures were more international in style), and by the early 1850s France too had her own schools of construction.

But meanwhile at least one major railway—from Paris through Rouen to le Havre—was to be built by a man trained under Stephenson. Even the stations on this originally looked English, though they were long ago smothered in French dressing. The great brick viaducts, however, still stride across the Ouest; and it is because of this link, more than any other in a thousand years of common history, that Normandy can almost be mistaken for a rather eccentric corner of England.

Fifteen years earlier, however, industrial Lancashire was coping with a surprise. The Liverpool and Manchester line, varied though its commerce was expected to be, had been built primarily as a freight carrier. But such was the pressure on it of the lucrative passenger trade that manufacturers were soon being told that engines could not be spared to cope with their goods. Not only the gentry but the lower orders were taking advantage of the speed of the steam locomotive to shuttle between the two cities.

The moral of this development was not missed by railways such as the Stockton and Darlington, which hastened to switch over to steam for their passenger traffic. Elsewhere, long-established tramways such as the Mansfield and Pinxton in Derbyshire re-equipped themselves for the new age, not merely introducing locomotives but making the engineering improvements which this step implied such as strengthening roadbeds and easing gradients and curves. In the early 1830s, too, a new generation of lines made their appearance—the St Helen's and Runcorn Gap, the steep Dundee and Newtyle which was the first railway in central Scotland, the Llanelly, the Glasgow and Garnkirk, the Bolton and Leigh which began an important network, and the twenty-mile Leeds and Selby which connected the former city with a deep-water channel. Though these typically used locomotives and carried passengers they also employed stone sleepers, as did the Bodmin and Wadebridge where the blocks remained in place on what was officially a normal LSWR branch until almost the end of the century.

Two others in this age-group of lines are of special interest. The

Leicester and Swannington railway of 1833 was sponsored by another Quaker, John Ellis, and was offered to George Stephenson; but the pioneer was busy and beginning to age (it is typical of these self-punishing giants that they were wearing out by their early fifties) and so passed the commission on to his son Robert, whose first independent work it became. The line had two inclines, one powered and one self-acting, at Swannington and Bagworth respectively: Stephenson's piston-valved engine at the former saw service until 1947, and at Bagworth the brake-house still survives. There was also a substantial tunnel at Glenfield, Leicestershire, whose ventilating shafts rise through private gardens and whose west portal is of particular interest. The line was laid on a combination of stone blocks and transverse and longitudinal sleepers.

The other unusual line was London's first, and significantly enough was a commuter railway. As opened in 1836 (against strong opposition, but to the music of bandsmen dressed as Beefeaters seated on top of the coaches), it ran from Bermondsey to Deptford; and it was remarkable in that, following the recommendations of another of those retired military engineers who provided a supplement to the Stephensonians, it was built on a viaduct throughout its four miles. This viaduct consisted—as it still consists, carrying very heavy traffic—of 878 stock-brick arches interrupted by some fine Doric bridges by George Landmann. The former were to provide a prototype for many railways in south London, but they no longer march across the pastoral landscape, dotted with cows and hedgerows, which led the company into the experiment of building elegant cottages under some of them.

Two years later this line, which was built partly to test the market, but also with dreams of a triumphal boulevard to the sea, was extended to a terminus south of London Bridge which became the nucleus of the capital's first great station, and in the country direction it achieved its objective at the desirable riverside village of Greenwich. Originally it was laid on stone blocks, but these were replaced by timber sleepers in the interests of quiet running.

Such is an outline of the railways of Britain opened between 1830 and 1836, a period in which Ireland, too, saw its first lines appear. This time-bracket coincides almost exactly with the reign of Sailor William (or, if one prefers, Silly Billy), and marks a suitable interregnum between the Georgian era of balance, proportion and gentle progress and the fiendish energy of that Victorian age which (at least in Britain) was to prove the most constructively dynamic the world has ever known.

These last lines to be built under the Hanoverian kings not only shared such characteristics as the use of stone blocks and fish-bellied rails but were still essentially short-haul branches built to serve local needs. Railways had now passed through four eras: the pre-technological years down to about 1600, the two phases of the early industrial revolution (which were themselves separated by the massive works built south of the Tyne around 1700), and the important transition period between 1825 and 1835. But they could not be called mature so long as few lines had a mileage measured even in double figures, let alone one approaching the thirty or forty miles of the Stockton and Darlington or the Liverpool and Manchester which itself fell short of a true trunk-route distance.

After the building of the second of these predecessors, as after that of the first, a lustrum of planning seemed needed before men took another step forwards. But then the age of experiment ended and the railways of Britain finally entered into their manhood.

The First Trunk Routes

Charles Lee defined the four characteristics of the true railway as
being a specialised track, the acceptance of public traffic, the carriage
of passengers, and mechanical traction. Quoting this list, Michael
Robbins added as a fifth criterion a measure of public control. But
it has already been suggested in this book that one more test of a
railway is that it should be of national rather than local importance.
This desideratum cannot be reduced to a matter of simple arithme-
tic; but at the least it implies the existence of lines more than fifty
miles in length and uniting different counties.

It is hence hard to agree with the late Canon Roger Lloyd that
the railway arrived suddenly and fully-formed. On the contrary it
was a rope woven of at least five strands, of which every strand so
depended on the others that passenger traffic would not be attracted
without the speed of locomotion and the locomotive itself needed reliable rails to carry it. The rate of weaving, too, was at one time
speeded by an internally generated impetus and at another slowed
by external events such as the Napoleonic wars.

These last (it should be interjected) delayed progress in civil
engineering more than they affected the mechanical department: in
fact, by inflating the price of fodder they actually speeded locomotive
development. But it is doubtful if they put back by much that final
stage in which such strong but short ropes as the Liverpool and Manchester were extended into cables which bound together the kingdom.

By 1830 the capital for such trunk routes had again become
available. Experience endorsed their profitability as well as their
usefulness, and engineering techniques were more than adequate
for their construction. But it seemed as if their projectors needed to
see a full-blooded railway in action before they were ready to go
ahead, for it was only after the Reform Bill riots and the further recession of the early 1830s that the first Bills for the construction of

long-distance railroads were presented to parliament. From 1833 onwards, however, such Bills were passed at close intervals. Their advocacy was often entrusted to the chief engineers appointed for the various lines, so that Robert Stephenson in particular had to add forensic skill to his other talents and civil engineers began to set up their head offices in the Westminster area.

Outwardly the years just before the midpoint of the decade were quiet ones from the railway viewpoint, but in fact they were fully occupied by the immense work of finalising routes and of detailed surveying. To consider the political considerations of routing in any detail would be outside the scope of this book; but it should be noted that, although the canal age had formed precedents for land acquisition and had set public before private interests so far as to introduce a mechanism of compulsory purchase, the early railway companies often had still to grind down or buy out recalcitrant landlords. In the event, though, routes had only rarely to be altered on pressure from influential landowners, exceptions being the successful opposition to the GWR of the Eton College authorities with their vision of an invasion of French governesses and to the Birmingham line of certain Vale of Aylesbury landowners. Far more often towns were anxious to attract the new transport medium, as (contrary to a long-enduring legend, and in the end without avail) was Northampton.

Similarly the menace to pioneering surveyors from farmers' pitchforks and gamekeepers' shotguns has been somewhat exaggerated in railway lore. There *were* incidents of this sort, particularly in the days when popular opposition to the Liverpool and Manchester was stirred up by a press campaign. But probably more blood was to be shed in the shennanigans between representatives of rival lines than ever was by the resistance of feudal squires, and the main initial opposition to be overcome was that of turnpike and canal proprietors.

Where the surveyor suffered most was through the soles of his feet; for the country he traversed, frequently in bitter weather, was often unsuited for even horseback travel. Robert Stephenson, for instance, reckoned that before his job was done he had perambulated the whole distance from London to Birmingham fifteen times over between twelve-hour bouts of non-stop dictation, and Joseph Locke often walked thirty miles a day inspecting as he went. Nights spent at fleabitten inns were little more restful, a problem which Brunel solved by equipping himself with a specially built caravan—the furnishings of which sound remarkably like those of the vehicle used by Field-Marshal Montgomery just over a century later, with the exception that Brunel gave generous space to the storage of his cigars.

The main task, however, was the assembly of the labour forces of many thousands of men needed for making each of these great lines. Again the canal builders had shown the way, and indeed some of the 'navigators' who had built the final canals in the 1820s survived to form a nucleus for the gangs of the railway age. But if the problems of recruiting, victualling, housing and disciplining these men were not new in kind, they were so in scale.

The railway navvies themselves swept across the face of an England where little had changed since the first Enclosures, and were often —like the Chartists, with whom they shared an image of destructive violence—feared like a hostile army. In a fashion they have passed into tradition, though they never created their own proud songs and legends as their American counterparts were to do thirty years later and it was indeed noted that they worked in total silence. Possibly their national background was too heterogeneous for a genuine folk-lore to emerge, for this labour force was *not* predominantly Irish and it is generally overlooked that the first phase of railway building was complete before the potato famine of the mid-1840s. The Irish were there, certainly; but so in equal numbers were the Scots and an English contingent drawn mainly from the Pennine counties but with an infusion of Lincolnshire ditchers and Cornish miners. Furthermore, the navvies proper formed only the contractor's élite of strength and skill, a permanent force which was supplemented by casual labour recruited *en route* and generally unwilling to operate far from its home ground.

Terry Coleman's recent book on *The Railway Navvies* assembles virtually all of the little accessible information about these literally nameless men with their moleskin suits and bright kerchiefs, their communal sex lives, their ration of two pounds of beef a day, their atrocious living conditions (according to Coleman a pub called 'The Shant' surviving at Four Marks, Hampshire, was converted from one of the shacks where a score or so of navvies and families shared a single room), and their bloodthirsty fights. Receiving their pay (of about £1 a week, or some three times a farm labourer's wage) at monthly intervals, they then went on those three-day 'randies' which led an observer to calculate that, for every mile of British railway laid, £1,000 was expended on liquor. At such times, in particular, the death-rate from disease and accidents—many of the latter being caused by the navvies' carelessness or bravado—was augmented by the lethal battles of the aggressive Scots against the comparatively pacific Irish, and of the English between themselves. The navvies loved a good funeral, and saw to it that these were not infrequent.

Of the aspect of their life which particularly concerns this book, the technical devices used, there is little to be said. The navvies' only real aid was black powder: used to form cuttings as well as tunnels, this was employed to shift not merely rock but sandstone, oolite, hard chalk and even clays. Tunnelling proceeded by the sequence of dig-out, shore-up and (if necessary) line familiar to seventeenth-century miners, except that railway tunnels were so spacious that fifty men could work on one face and were usually also shallow enough to allow vertical access shafts to be sunk every hundred yards or so. In soft soils a simple shield rig was employed, such as Marc Brunel had devised in 1818 after noting the habits of shipworms and had used on his Thames tunnel.

Embanking followed the methods known to Telford, with the addition that an ingenious (and hazardous) method was devised for automatically tipping the 'muck' end-on from the head of a bank: since soil mechanics were little understood, such earthworks were not often higher than some 60 feet. For cuttings, which in soft soil were rarely of more than a similar depth, the first stage was to dig a narrow 'gullet', this being later widened by the technique familiar from contemporary illustrations to anyone who has dipped into railway literature: plank 'runs' were arranged up the sides of the cutting, and on these the navvies guided wheelbarrows. The motive power was supplied, *via* a rope and pulley, by a well-trained horse pacing a walk at the top.

In slippery weather this was hazardous work; but tumbles were accepted as routine, limbs and lives were cheap, and in fact the navvies angrily resisted the attempts of one engineer to introduce a safer method and broke up his device like good Luddites. A few primitive mechanical aids such as cranes, boring rigs and piledrivers, all worked by horse treadmills or 'gins', were used in appropriate situations, but the only important innovation of railway over canal civil engineering was the natural enough one of building temporary light tramways (worked as soon as possible by steam locomotives) as work progressed: these simplified the transport of spoil prior to the construction of a 'permanent way'. Wherever possible cut-and-fill programmes were followed; but topographers can still discover many 'spoil banks' left near cuttings and tunnels, some of these being marked as such on ordnance maps as at Potters Bar. Less often, land beside embanked tracks displays the 'borrow pits' of the corresponding operation.

There is no memorial, though, to those manufacturers of wheelbarrows who must have grown rich in this age; and the only spades to achieve immortality are the hideously engraved ceremonial ones used for cutting first sods, one of which ignominiously folded in two on the occasion of its sole use. Yet the spade it was, in all its hundred-odd local and even religious variants, which conquered Britain in this age which still relied more on manpower than even animal-power. It has been calculated that the John Henrys of British railroad construction—the men who carried their own tools and barrows on their backs from site to site—lifted their muck at an average rate of over a hundred foot-tons a day.

These limited technical resources, however, were backed by a characteristically Victorian belief that nothing was impossible. And so at last, in the boom year of 1836 which marked the eve of the Queen's reign, the railway age reached its long-delayed point of take-off and steam and iron were together ready to conquer Britain. An interesting point developed by Hamilton Ellis is that this conjunction was not inevitable: the canals rejected steam power on the grounds of the damage which wash would cause to their banks (and in so doing hastened their end), but had the roads been a little more even and a little less turnpike-minded the history of land transport might have taken a different turn and Telford's plan to lay the Holyhead road with a reserved track for steam road vehicles be today regarded as an historic landmark. As it was, over half a dozen major railway schemes were begun in or about 1836 (during which year a hundred more Bills were passed), and were to result in lines opened within a year or two either side of the end of the decade.

These lines cannot be discussed in strict chronological order: too much was happening too fast. Nor do the great trunk routes fall into a tidy topographical pattern, for outside the mind of William James national railway building never knew even the attempt at planning which marked the start of the canal age under Brindley. Instead the new era began, as that earlier period had ended, in the individualistic confusion which was to mark the whole history of railway building. But certainly no British railway, and perhaps none in the world, was to be of greater importance than that which realised James's own dream of uniting London and Birmingham.

In 1830—four years after the failure of James's scheme, and with the Liverpool and Manchester now in operation—surveys for a more direct route of 112 miles were carried out by the Rennies and Rastrick as well as by the Stephensons: these latter, as a result of their earlier successes, were eventually appointed. As on the Leicester and Swannington George carried out the preliminary negotiations, which included some tough parliamentary committee-work following an initial rejection of the scheme. But as on the Leicestershire line too the engineering was left in the hands of his son.

In 1833, when the Birmingham project was finally approved, Robert was thirty. By modern standards he was a child to be put in charge of such an unprecedented work before he had proved himself even on the short Leicester railway. But although George Stephenson is often regarded as the father of railways, and although he certainly vastly helped them to emerge from their tramway prehistory, he was in fact a mining and general engineer. It was Robert, a generation younger and almost exactly the contemporary of other leading pioneers, who has the greater claim to be considered the founder of railway civil engineering. He could also be considered the more talented of the pair, though he of course stood on the shoulders of his self-made father. The men were perhaps as important as each other in the field of locomotive engineering—which is to say, of very great importance.

Construction began at the London end in 1836 with John Birkinshaw as resident engineer, one of the first major tasks being to cross the Chiltern chalk by way of a cutting some $2\frac{1}{2}$ miles long and up to 60ft deep. Boxmoor, with its fine embankment, was reached in July 1837 and Tring in October of that year. Then in the April of 1838 a 'mean alehouse', only recently destroyed, near Denbigh Hall on Watling Street served as a temporary terminus before the line was pressed on by way of another great cutting of $1\frac{1}{2}$ miles (this time 70ft deep, and through limestone supported on waterlogged clay) to Roade in Northamptonshire and so to Wolverton.

Meanwhile work had progressed southward from Birmingham through Coventry to Rugby, which town was reached in September 1838. It is interesting to note how closely the southern section follows the route of William Jessop's Grand Junction (later, Grand Union) canal built forty years earlier; but to say this does not belittle Stephenson's surveying talents but rather stresses that there is often only one 'best' course between two terminal points and that the canal builders knew how to compute it. Elsewhere in Britain canals were to be closely shouldered by their bitter rivals the railways; and when, another century or more on, motorways began to be scratched across the countryside they frequently had to be fitted into the same narrow corridors. Across Buckinghamshire, Bedfordshire and Northamptonshire, for instance, the Roman A5, medieval roads, the canal, the main-line railway and the M1 parallel each other or cross and recross, each route often being in sight of one or two others even when all are far from a town.

So far eight tunnels had been built on the London and Birmingham line, though only two were more than a few hundred yards in length. Those at the southern end (such as Primrose Hill) had mildly decorated portals, though the style used was too chaste to be given a stylistic name and both Stephensons had enough of the puritan in their make-up to avoid ornamentation unless there was a good reason for it. A happy structural detail, however, appeared on those tunnels whose portals were strengthened by being returned parallel to the tracks in graceful falling curves.

Such 'horizontal arches' had been typical of Rennie's canal aqueducts (where they accorded pleasantly with the lovely 'change-line' bridges) but were not to become common in railway engineering. Stephenson's use of the strong, purple 'engineers' bricks called 'Staffordshire blues', however, was to characterise not only lines associated with the London and Birmingham but also many others in the south and Midlands where building stone was not available. They were particularly employed in an innovation of railway engineering, the arched retaining walls used to hold back the sides of steep cuttings.

Meanwhile a last link remained to be forged. This took the line through the Northamptonshire oolite by way of a 1¼-mile tunnel;

and though some of the earlier tunnels had presented difficulties (at Watford shifting ground had been struck and lives lost, the lining at Primrose Hill had cracked under its load of London clay, and in almost every case invert arches had had to be added as afterthoughts), this provided the major hazard of the route. Ironically, the parallel canal tunnel had proceeded fairly smoothly and Jessop's major problem had been the spanning of the Ouse—which afforded Stephenson no other difficulties than the destructive opposition of canal proprietors, if one discounts the curious fact that some of the Northamptonshire shale proved to be spontaneously inflammable.

But by ill fortune Stephenson struck quicksands, missed by his trial bores, in the heart of the Kilsby tunnel. His contractor died of the shock; and after consulting his father Robert took over in person, using not only a labour force of some 1,300 men and 20 horses but also (when a lateral drift had failed to drain the tunnel) thirteen steam pumps together capable of lifting 2,000 gallons a minute up some 150 feet. This event marks the first substantial use of powered equipment in railway civil engineering.

Thus, after a delay of eight months, the loss of twenty-six lives, and labours extended over more than four years, the tunnel was completed. Considered by L. T. C. Rolt one of the great epics of engineering history, it had cost some £250,000 a mile as against well under £50,000 for the line as a whole, and had progressed at the rate of a yard a day as compared to the average of over 100 yards. But in 1838 trains could at last run through from Euston Square in London to Birmingham's Curzon Street terminus.

These stations are described later. Meanwhile, it should be noted that a number of branches to the London and Birmingham had been started. Thus, perhaps Britain's first passenger-carrying railway branch—again following a canal route—was that which ran down to Aylesbury and was completed in 1839. This is now closed; but the main-line tunnel of course survives, with the feature that two of its twenty-five working shafts were provided with castellated vents and finished off at a diameter of 60ft. Far larger than the tunnel itself—as broad, in fact, as a modern passenger coach is long—these still on a sunny day cast dramatic pools of light down their height of 100ft or so. Their immense bore, never to be repeated, suggests that Stephenson overestimated either his passengers' need for fresh air or the 'piston effect' of resistance to a train entering a long tunnel at high speed.

The completion of the line (which, like all those built at the period, was double-tracked throughout) was suitably celebrated with an engineers' dinner lasting for over fourteen hours. Some contemporary writers compared the building of the 'Brummagem' to that of the Pyramids, others to that of the Great Wall of China, while others again calculated that it was the greatest work ever accomplished by human hands. But what is certain is that the London and Birmingham railway remains very typical of early construction in its easy overall gradient of 1 in 330.

There was only one incline, leading from Camden Town down to the Euston terminus at 1 in 75; but here cable haulage was necessary for the first ten years or so of the line. It may be mentioned that the approach to a city terminal was in general to prove the last redoubt of the cableway for passenger trains, with a 1 in 48 incline operating at Cowlairs, Glasgow, until 1908. On mineral lines it was, of course, to endure far longer, for entire coal trains in Wales were cable-hauled up one bank into the British Railways era and on Tyneside several inclines still work with 'cuts' of a few wagons at a time.

Robert Stephenson also used the conventional, chaired, fish-bellied rail of the period. He did, however, advance from his father's practice on the Liverpool and Manchester in taking heavy wooden sleepers laid transversely as his norm. In cuttings he adhered to the stone blocks of the tramway age, presumably through fear of damp and rot. It was to be a few more years yet before these over-cautious north country engineers were convinced of the durability of well treated and elastic hardwood under almost any conditions once sound drainage had been provided.

To skim up to Birmingham in today's electrified expresses is not, perhaps, a great scenic experience. But for anyone with a sense of the railway past there is scarcely an uneventful five minutes as cutting, viaduct and tunnel define the gentle contours and reflect the changing geology of the south Midland plains. And for virtually every inch of the way one is passing over Robert Stephenson's own roadbed.

For it was not only in their easy gradients (which are explicable on the grounds of a mistrust of locomotive power and adhesion) that the pioneering engineers built for the future. The radii of their curves were rarely less than a mile, which proved adequate for speeds far greater (and rigid wheelbases longer) than those they knew; and they also worked out the accompanying problems of the 'cant' or 'superelevation' of the outside rail necessary for a train to round a bend safely and of the need to lead gently into such curves. Even Stephenson's water-tower at Blisworth, believed to be the first of its kind in the world, outlived the age of steam and was shamefully dismantled after 130 years of continual use.

In some directions—for example, the comparatively rare level crossings built to a stoutness later found inappropriate—these exaggerated factors of safety landed the nation's railways with an unnecessary capital load. It should also be noted that in order to accommodate landowners Britain's railways (following her canals) had to build minor bridges at ridiculously close intervals: even in open country the average was about one every half-mile, while in the London suburbs Stephenson was ordered to provide under-bridges to allow for future urban growth. Many of these, too, were of the costly skewed form for which timber models had to be built. But caution as to loadings was to prove of a value which the pioneers could have foreseen only subconsciously—a fact shown by the London and Birmingham more clearly than by any other railway.

One's first reaction on glancing at a Bourne or Ackermann print of the railways of the 1830s and early 1840s is of the disparity between their anatomy and their physiology. *Here* is a tremendous viaduct built to outlive the Caesars and *there*, crossing it, is a toy train with a high-chimneyed puff-puff engine pulling a few stage-coaches and open trucks painted in jolly, playground colours of primrose and pastel blue. Except when a special with up to sixty coaches and five locomotives was running, the apparent overbuilding must even in its time have seemed almost ludicrous. Yet the country had good reason to be grateful to Robert Stephenson as the decades passed and the London and Birmingham, fulfilling its promise as part of Britain's medial rail system, bore ever heavier loads and eventually became part of a major modernisation project.

The propaganda of that later age referred to 'Britain's new railway'. Certainly the road was relaid throughout, clearances had to be eased and overbridges raised, and the precaution was taken of strengthening many underbridges too. (Thus, two of the original metal spans in the Wolverton area were encased in concrete. That which crosses the A5 road in Buckinghamshire, however, is on a gentle skew and remains in its original state with stone abutments supporting an iron deck.) But not only most of Stephenson's single-span masonry underbridges but his long viaducts too were found fit, after more than a century and a quarter of increasingly heavy service with only routine maintenance, to bear unstrengthened today's high-speed electrified traffic.

The contrast between conditions then and now are summarised in the following table. The figures are round ones, but the moral is clear:

	1840	1970
No. of through passenger trains per day	24	42
Speed (average, mph)	20	75
Speed (maximum, mph)	40	100
Locomotive weight (standard, tons)	14	81
Carriage weight (standard, tons)	4	32
Passenger train weight (average, tons)	60	350
Freight train weight (maximum, tons)	250	1,000

The unplanned form of Britain's railway system has already been noted: the first lines were simply opened, as the tramways had been, where there appeared a local need for them. But with Liverpool and Manchester already rail-linked the desirability of uniting both to the London and Birmingham line was obvious, especially since the two systems had the same gauge and other characteristics. And in fact, from the granting of parliamentary sanction for the London and Birmingham onwards, work had proceeded simultaneously on a line projected even earlier to run north from Birmingham to Warrington in Lancashire. From there a short branch already existed to Newton-le-Willows, near the midpoint of the Liverpool and Manchester line.

Originally known as the Grand Junction railway, this eighty-mile link worked in conjunction with the London and Birmingham to provide a through service from the capital to industrial Lancashire. At first, though, passengers had to change at Birmingham from Curzon Street to the (now vanished) Grand Junction terminus. And there were other differences of opinion between the two systems.

Back in the days of the construction of the Liverpool and Manchester, one of George Stephenson's assistants had gone badly astray in setting out a tunnel: had his plans been followed, indeed, the headings would never have met. Another young assistant—Joseph Locke, the son of Stephenson's closest friend, who had worked on the

Canterbury and Whitstable line—had been called in to put things right, and had made a good impression on the influential 'Liverpool party' of financiers. Hence, after Locke had split with Stephenson and set up on his own, he was appointed engineer on the Grand Junction.

While in no way reverting to the deviousness of the light tramways, and while sharing the contemporary horror of long, steep gradients, Locke appears to have been more prepared than the Stephensons to compromise with his topography. Hamilton Ellis has summarised the two philosophies as 'up and round' rather than 'straight through', and though this may be an oversimplification Locke often went to some pains to avoid even the shortest and shallowest of tunnels. On the Grand Junction, however, he was for the most part faced with the easy terrain of Cheshire where the main hazard to the railway engineer was one which only showed itself over decades, the subsidence caused by salt extraction; and hence he could work for miles on end to traces almost as straight as a Roman highway.

His route ran from Birmingham to Wolverhampton (though not by the present main line), and continued north by way of some substantial works and a short peak at 1 in 176 near Penkridge to Stafford. It then passed a spot 'near Crewe Hall' before sweeping towards its conclusion over a fine viaduct of twenty sandstone arches across the Weaver; and there was a final twelve-arched viaduct over the Mersey and its lateral canal. Work proceeded smoothly, and the line was opened throughout a year before the London and Birmingham. It had been built at a third of the cost per mile of this, thanks partly to a low figure for the land acquisition costs which then accounted for about a quarter of the total expenditure on a new railway.

Locke used timber sleepers spaced at intervals of some 30 inches on most of his system, and also made an important advance by perfecting (if he did not invent) a double-headed form of rail whose section took the form of a dumb-bell. This was keyed into chairs with hardwood wedges, the centre lines of each pair of rails sloping slightly inwards. The initial idea had been that after one face had become worn down the rail could be turned over, but this did not work out in practice because the lower faces became dented by impact with the chairs. Hence the section was modified to that of an

6 *Top* Blisworth,
7 *Bottom* Washington

GE OVER THE RIVER MERSEY AND CANAL

LONGITUDINAL SECTION OF CANAL ABUTMENT

MASONRY BRIDGES

8 *Top* London and Greenwich railway
9 *Middle* Grand Junction railway
10 *Lower* Great Western railway
11 *Bottom* Midland Counties railway

inverted figure-eight, with the upper bulge substantially larger than the lower. With one major improvement, and a few minor ones in retaining devices, this 'bull-head' rail was to serve as standard on Britain's railways for over a century.

Another far-sighted move of Locke's was to entrust much of his work to a single and adequately-financed contractor. Here even Robert Stephenson, though a better organiser and delegator than his father, had often gone astray, allowing his work to be let out in job lots often little more than a mile long. Some of the firms handling the 'Brummagem's' labour force of between 10,000 and 20,000 men were of the high repute of the Cubitts who worked at the London end. Others were 'Geordie' friends from the old days, now setting up on their own account: these brought mining knowhow and a sense of teamwork to the railway scene, but were subliterate and often went wrong in their costing. And yet other contractors were village enterprises whose skills and integrity were as questionable as their resources, who might rely for their profits mainly on the operation of the loathsome 'truck' system, and who in turn had to subcontract and sub-subcontract down to the level of the ganger working with a dozen mates.

Locke not only avoided such pitfalls: he was unusually lucky, even in an age when men of talent seemed to press forward as fast as the demands of railway building called for them, in inheriting from Stephenson a contractor of the calibre of Thomas Brassey. For this great engineer, who had once worked for Telford, was to go on to build nearly 2,000 miles of track in Britain alone, to collaborate with Locke in France, to lay the foundations of railway systems in several other western European countries, and to construct the (financially disastrous) Canadian Grand Trunk railway, all without ever sacrificing the respect and even love of his workmen as well as his clients. Men with Brassey's moral and financial resources were not only unlikely to go bankrupt as a result of comparatively minor misadventures as did ten of the thirty main contractors on the 'Brummagem': they also made possible more accurate costing and quantity surveying.

One of the factors common to the first trunk routes, for instance, is that their estimates generally represented only about half the final expense. Considering the achievements of our own age of computers, critical-path surveys and continual inflation, when the allowance for a new aircraft can easily be overrun by a factor of ten, this may seem remarkably accurate forecasting. But the errors attracted adverse comment in that more vigilant era.

In charge of the fortunes of up to 100,000 men at a time (or about the same number as were needed to build the Great Pyramid), Brassey was often called the commander of a private army. It was pointed out, too, that his budget was greater than that of many a minor state. But he was not unique here, for Morton Peto became a contractor on almost as huge a scale—though he finally bankrupted himself in the crash of 1866 which also ruined the Quaker railway financier Gurney, whereas Brassey died a very rich man. Between them Brassey and Peto were to do much to recoup the results of military idiocy in the Crimea by building the thirty-mile Balaclava railway in a few months—a superbly planned private operation, as even the generals admitted, and one which not only founded army transport engineering but provided a strong counter argument to those early socialists who thought that Britain's railways would have been better built under state control.

Meanwhile, though a plan dating back to Telford's time to link Glasgow and Edinburgh was to be completed in 1842 by way of an almost level track of over forty miles incorporating Grainger and Miller's fine viaduct of thirty-six arches at Clifton Hall, no line had yet spanned the width of England. But while Robert Stephenson was completing the 'Brummagem' his father was at work on a forty-mile line which was to cross the Pennines and link Manchester and its manufactures to the populous West Riding of Yorkshire.

Three canals had essayed this watershed, but the rather indirect Leeds and Manchester railway (later the spine of the Lancashire and Yorkshire system) took what appeared the easiest route of all and had been surveyed as far back as 1824. However, the Summit tunnel at Littleborough, though comparatively shallow, was over $1\frac{1}{2}$ miles long and hence the lengthiest railway tunnel in the world at its opening; and it gave Stephenson and a thousand navvies considerable trouble and over four years of work.

A long viaduct, too, had to be built through what is now the centre of Todmorden, just south of which town the original 'bow-string' iron bridge still carries rail traffic skew-wise across the disused Rochdale canal. As had been the case at Birmingham, and was to become unnecessarily common, a passenger terminus was built

in Manchester independent of the existing one and was not connected
to it until 1844: due to changes in railway politics the eastern
terminal was shifted to a junction near Normanton. The line was
eventually opened, on stone blocks throughout, in 1841.

Before the Leeds and Manchester railway was complete, George
Stephenson had moved on to a cluster of neighbouring lines. But
before this book itself passes to these—or to the three fine but very
different railways taking shape in the south—another early transverse route should be mentioned, even though this was far to the
north and (unexpectedly and contrary to some accounts) owed little
to the Stephensons. It was in fact first surveyed by that believer in
cable traction, Benjamin Thompson; but the engineer put in charge
of construction was Francis Giles, a pupil of Rennie's who had
formerly specialised in harbour works.

As far back as 1829 an Act had been passed to prolong the existing
lines from Newcastle and Gateshead to the Tyne coalfields across the
highlands by way of Hexham to Carlisle: the track passed only a few
hundred yards from the house at Wylam where Robert was born
when a timber tramway ran by the door—a house which, like the
Stephensons' later one at Killingworth, is still preserved. This
Newcastle and Carlisle railway, with a main line of over sixty miles
following the Tyne for much of its length, was completed by 1839.
Perhaps because at the eastern end it connected with a host of
colliery tramways built in the 1830s, and perhaps too through the
ready availability of stone, it reverted to the old Stephensonian/
Northumbrian type in using stone-block sleepers as well as fish-
bellied rails throughout: it was also intended for horse traction,
though it carried passengers from the start and in its engineering
was indisputably a railway. The countryside was wild rather than
difficult and Giles was hence able to avoid major tunnels and cuttings.
He built, though, an impressive viaduct near Wetheral and a large
number of bridges over the becks.

Even apart from the stations mentioned later the Newcastle
and Carlisle remains today of unusual interest. For nearly a decade
it was detached from the rest of the country's railway system, it never
had cause to be served by through trains, and together with its
Alston branch (built twenty years later, but appearing contemporary
with the main line and happily still open) it still affords the least-
changed example of all early railways if only because its traffic has

never either increased nor decreased dramatically. As it sweeps across the ragged borderlands close to those forts where the Romans ground out their ruts 1,500 years before it still appears *sui generis*, neither branch line nor true trunk route.

At the conclusion of this chapter it is perhaps worth asking how the average Englishman, neither engineer nor *entrepreneur*, neither squire nor smallholder nor turnpike operator nor aesthete, viewed such an invasion of Britain by titans. Romantic of the swiftly-supplanted stage coach age though he was, Charles Dickens gave vivid pictures of several facets of the railway era, and of particular interest is his sketch in *Dombey and Son* of the temporary chaos as the Primrose Hill tunnel bored its way below what was then the fringe of London. But for the most part the trunk routes crossed virgin country; and there a typical observer was George Eliot, with her calm acceptance of the change which the railways brought to her views of the Midland plains.

Railway engineering did not blend into the landscape as naturally as had most of the contour-linked canal works. It made its presence felt, and the lady of title who graciously permitted her Chiltern estates to be bisected by the Birmingham line on the grounds that they were already 'gashed' by a canal can be suspected of being ruled more by financial than by aesthetic considerations. But proud as the new structures were they were never arrogant, and they made enough use of local materials to be accepted by all but the most hidebound once their novelty had passed.

In Britain, at least, not even the Romans had built on the scale of Edge Hill or Tring. But the acute Fanny Kemble had noted ferns growing at Olive Mount even before the railway there was opened, and forty years later Thomas Hardy was to remark on how swiftly nature reclaimed the land. A few seasons of sun and rain and snow revolved: a brick or masonry surface was eroded a little by the elements and stained a little by locomotive smoke: and within five years the viaduct, cutting or station was at one with the manor house, church and inn.

CHAPTER FIVE

From London to the Sea

The lines opened in the 1830s spanned generally increasing lengths—30 miles for the Liverpool and Manchester, 60 for the Newcastle and Carlisle, 80 for the Grand Junction and over 110 for the London and Birmingham. With the last of these the trunk railway became a reality; and in many ways its form was as firmly established as its existence.

Much of this form was derived from the characteristics of the steam railway itself as surely as the differing traces of a neolithic, Roman or medieval road derived from the capabilities and needs of the men and beasts who used them, or the route of a canal depended on its need to gather water and contain it at constant levels. Thus, the primary virtue of a rail system was that it enormously reduced friction and so made possible the haulage at speed of heavy loads behind a comparatively low-powered engine. But this advantage over road vehicles diminished fast once the route left the level and gravity became a major factor, since the prime mover hauling loads other than over a specially prepared track necessarily had to have power in hand and could always—through a change of gearing or simply a natural slackening of speed—convert antifrictional to antigravitational effort.

It is true that the earliest railway engineers were (as is shown by the experiments of Blenkinsop and Murray and by the contemporary 'crawling' engines) unduly cautious about the adhesive properties of the conventional railway, and assumed that a locomotive's wheels would spin idly round long before it reached the point of stalling. It is true too that the next generation held on too long to the idea of the incline, and that the makers of the trunk routes themselves underestimated coming developments in locomotive power. But future generations were on the whole to be grateful for the mis-apprehension that any gradient steeper than 1 in 300 (which, coincidentally, was equivalent to the canal average of a lock about every three-quarters of a mile) should be looked at askance, and that 1 in 100 should be regarded as a limit to be resorted to only for lengths of a mile or two at the most.

For this was certainly a truer view of the nature of a railway than that of the half-informed parliamentarians who saw early trains taking short steep lengths on the run, and who went on to argue that money was being wasted on tunnels and viaducts and that whatever route an engineer built on he would end up with his energy in balance. Presumably on this theory of the 'undulating railway' not only the terminals but all intermediate stopping-places would be built at the same altitude, a principle which was to prove useful only on completely artificial lines such as underground systems.

The surveys of the railways of the middle 1830s hence had a sound theoretical basis. But the transverse wooden sleeper had not yet driven out the stone block, rail form was very far from standardised, and the choice of gauge could still be regarded as almost an arbitrary one; for whether or not the famous figure of 4ft 8½in had any roots in antiquity, the proximate cause for its adoption on the great passenger-carrying lines was simply that George Stephenson had inherited it from his Northumbrian wagonways and saw no reason to change it. Although there are references in early literature to both 4ft 8in and 4ft 9in, it is doubtful if these represent more than roundings-off, differing ideas as to datum lines (correct measurements being taken between the inside faces of the rails), or perhaps local differences in clearance.

In 1836, however, Parliament had feebly legislated in favour of a compulsory 4ft 8½in; and well before 1840 this was becoming considered the national standard, though even after Victorian times it was often (and confusingly) referred to as the 'narrow' gauge. It was proving reasonably convenient for freight carriage and comfortable for passengers, but *only* reasonably so. It represented a compromise between the narrower rail-settings which could worm round tight mountain bends and the broader ones which promised smoother and more economical running—but was it the ideal compromise?

By 1835 there were in Britain nearly five hundred miles of passenger-carrying railway track—virtually all planned by the Stephensons or their disciples—either already open or under construction and committed to a 4ft 8½in gauge. In addition, this gauge had taken roots in the USA (where, however, a range between 3ft and 8ft could be found even at the end of the century), and with benefits which only future ages were fully to appreciate was also established on the continent of Europe. But track-gauge was the most important parameter of a railway; for it controlled locomotive power, carriage capacity and permissible radii of curvature (and hence, respectively if indirectly, speed, comfort and the ability to penetrate rough country) as well as facets of railway technology as yet hardly thought of such as the layout of sleeping cars. It was hence arguable that 'the gauge question' should not have been left unchallenged at a figure derived from the hazards of north-country collieries and that one engineer, at least, was right to go back to fundamentals, to ask himself: 'Is this the best gauge?' and 'If not, is it too late to alter it?', and perhaps to answer these questions in the negative.

Meanwhile the prosperity of the port of Liverpool—a prosperity which had both given rise to and been augmented by the building of the Manchester railway—had caused concern to the merchant venturers of its more ancient rival, Bristol. These merchants decided not only to modernise their harbour installations but to exploit their greater proximity to London by building a railway to the capital. Legislation (which deliberately left open the question of gauge) went through parliament in the wake of that for the London and Birmingham line, and in 1835 a chief engineer was selected.

At that time Isambard Kingdom Brunel was twenty-nine, younger even than Robert Stephenson had been when he was appointed to the London and Birmingham. He had no railway background, and indeed no achievement of any kind behind him except for his work on the disaster-ridden Thames tunnel; and at the critical time he had been unemployed for some years. He had impressed influential Bristolians with his plans for the Clifton suspension bridge, but even so his name was only added as an after-thought to a short list of nonentities for the railway project. Yet, as happens so often and so mysteriously, the British system of uninformed committees and backroom manoeuvrings hit on an individualistic genius, one prepared to challenge the Northumbrian monopoly. And monopoly is not too strong a word, for at this time the Stephensons were mentally carving the country into slices and assigning projected standard-gauge trunk routes to their associates, with the west being promised (by a striking irony) to the Thomas Gooch whose rebellious brother Daniel was to become Brunel's disciple and locomotive adviser.

Brunel himself was the maverick who, right or wrong, was still necessary. The only railway pioneer to be quite independent of the north-country school, he also had a different social background; for whereas even Robert Stephenson was only removed by one generation from the coal-face, Brunel was the London-born son of a professional engineer who had come from a long line of prosperous Norman farmers. But above all he was an engineering radical working without preconceptions, so that when his great chance came he seized it by redesigning the whole structure of rail and railbed.

His solution, arrived at after he had travelled on the Liverpool and Manchester and found its accommodation cramped and its running uneven, is familiar to anyone with a slight knowledge of railway history. For his rail he chose the section resembling an inverted, wide-seriffed U or broad-brimmed hat known as the 'bridge' rail: this, he believed, gave the best possible strength/weight ratio. Furthermore its light section would help overcome another trouble, since with the hasty production of blooms welded-up from plates of puddled iron the heavy Birkinshaw rails seem to have declined in quality and were flaking or laminating and generally not living up to their early promise.

Even at the start these bridge rails weighed 55 lb/yd, and later versions reached 62 and even 75 lb. They were throughout supported on longitudinal bearers or 'baulks' of heavy pine faced with hardwood, and these were braced to each other by cross-ties which, where doubled tracks were laid, spanned both roads. Finally, these widely spaced 'transoms' were spiked to piles which descended vertically into the roadbed, so that the timberwork was anchored in three dimensions before it was rigidly ballasted and then planed down to accommodate the rails. The gauge was set at a nominal 7ft (and actual 7ft 0¼in.), seemingly an arbitrary figure but one which Isambard Brunel may have taken from a crane track built by his father in Chatham dockyard. He later said it should have been wider: to appreciate how wide it was, one can envisage an ordinary bed set crosswise sitting easily between the rails.

Brunel's road proved slightly cheaper in materials and main-

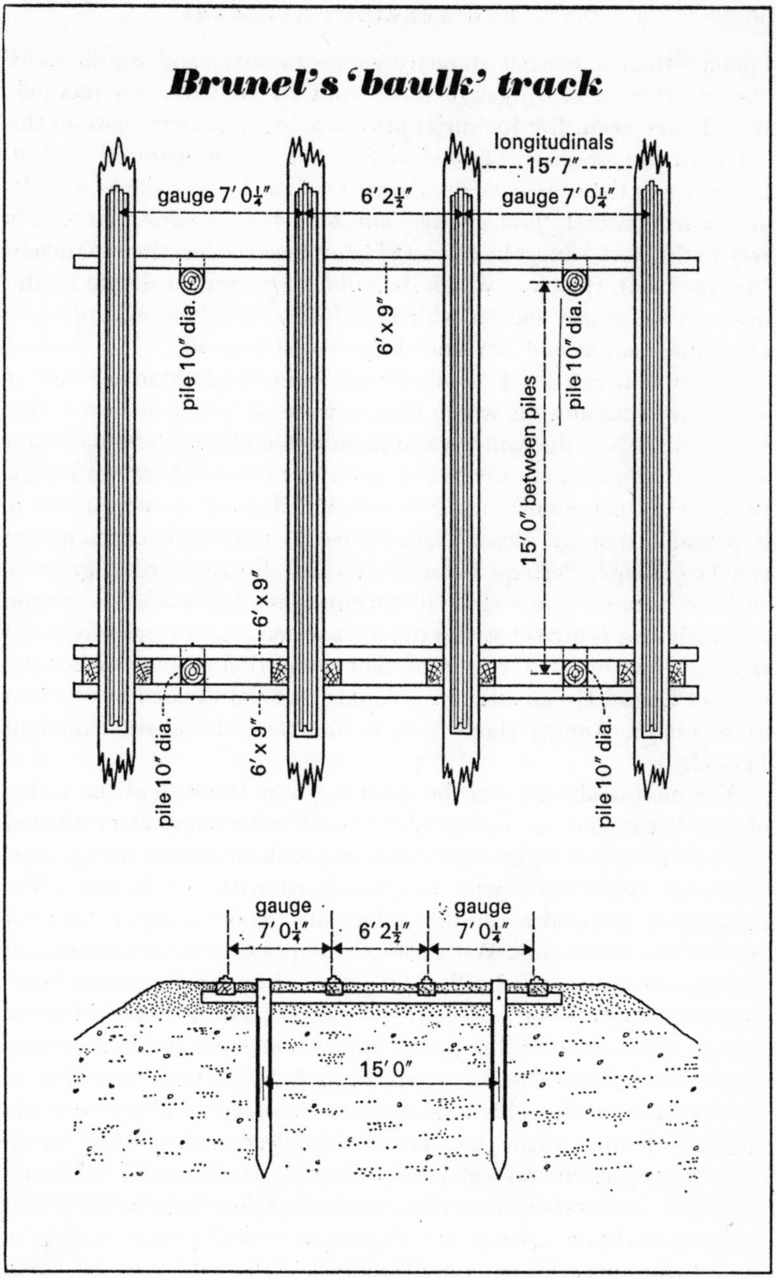

Brunel's 'baulk' track

longitudinals --- 15' 7" -----

gauge 7' 0¼"   6' 2½"   gauge 7' 0¼"

pile 10" dia.

6' x 9"

15' 0" between piles

pile 10" dia.

6' x 9"

pile 10" dia.

6' x 9"

pile 10" dia.

gauge 7' 0¼"   6' 2½"   gauge 7' 0¼"

15' 0"

Figure 2

tenance than a normal standard-gauge railway, and considerably
cheaper than a broad-gauge track built by conventional methods
would have been. But its initial price was high, in terms both of the
extra 4ft 6in or more of land which had to be acquired and of
labour costs. Other merits and demerits of the broad gauge—and, in
an English context, 'broad gauge' almost always implies Brunel's 7ft
way to the west—must be discussed later. For it is now time to turn to
the way itself, that way which its builder determined should be the
finest in the world and on which (as Robert Stephenson confessed)
only imagination and not knowledge could improve.

As built, it ran for 118 miles west from Paddington by way of
Wormwood Scrubs, on which dismal waste it passed within a few
hundred yards of the London and Birmingham line. There had been
a serious proposal not only to link the two railways there but to run
them in to a common London terminal, and had this materialised it
is probable that the Bristol railway would have had to accept the
standard gauge. Perhaps fortunately (for railways were to grow at
such an unforseen rate that the terminus would soon have become
an overloaded source of strife) the lines went their several ways; and
though a few years later an important connection was established the
dead of Kensal Green still sleep flanked but not divided by the two
great routes. Among those dead is numbered Isambard Kingdom
Brunel.

The next landmark was the spanning near Hanwell of the valley
of the Brent and its associated artificial waterways. Here Brunel
built a viaduct of eight stock-brick arches dedicated to his sponsor,
the Lord Wharncliffe who was associated with the Grand Allies
consortium founded a century before and whose arms can still be
seen on the south side. With both the depression of the arches and
the taper of the coupled pillars exaggerated by Stephensonian standards, this typical work of Brunel's is often, like his road bridge at
Clifton, referred to as 'Egyptian'. But whereas the architects of the
masonic temples of this time consciously imitated what they took to
be the lines of the pharaohs, it is probably true to say that the
engineer Brunel made his natural predilection for boldly exaggerated proportions respectable by adopting the name of a historic
style. His motivation was that which had led him to dream of
bridging the lower Thames in a single span, a Gothic desire to get the
utmost from his brick or stone. For Brunel (who had been let down

by one early metal bridge) mistrusted cast iron in structures and thought that Robert Stephenson was loading it too heavily in some of his underbridges. This opinion was shared, perhaps correctly, by Joseph Locke.

In June 1838 the line completed its first stage and reached Taplow in Buckinghamshire. The next challenge was the crossing of the Thames, which the navigation authorities insisted should be carried out in two spans at the most. Each arch would hence be 128ft across, while Brunel's rail level was little more than 24ft above the water. The result was a bridge after the iconoclastic engineer's own heart, one whose ratio of major to minor axis of the semi-ellipse was greater than any hitherto attempted; and when its centering was finally eased in the autumn of 1838 there were those who gleefully awaited its fall.

In fact, Brunel had agreed to leave the centering in place throughout the winter months as a precaution against frost-cracking; but others who watched the work trains crossing it believed that he was afraid to let the bridge bear even its own weight and were surprised when eventually a storm carried the timberwork down into the swollen Thames without any bricks following. This is a classic anecdote of railway history, and one wonders if Brunel was aware that about a century and a half earlier Christopher Wren had played a rather similar trick with his columns which did not *quite* reach the ceiling of the town hall of nearby Windsor—that historic town which could not make up its mind whether or not it wanted a railway and which, in the event, was served by a short branch carried across the river by an interesting early bridge.

By the summer of 1839 trains were running to Twyford. Meanwhile, in discouraging weather, Brunel had driven a cutting originally planned as a tunnel two miles long and 60ft deep through the chalk at Sonning. His difficulties there were almost contemporaneous with those which Robert Stephenson was facing at Kilsby, and like Stephenson he found his contractors going bankrupt and the completion of the work left in his own hands. But though the two men were, for all their differences, to remain on good terms for over twenty years they could not learn from each other now. Brunel had to break through alone, to reach Reading in the spring of 1840.

Here there was a parting of ways, for the traditional route followed by the coach road to Bath (and by the Kennet and Avon canal) ran south by way of Newbury and the Vale of Pewsey. Brunel, however, had decided to follow the Thames valley itself to the area of Abingdon, where the hamlet of Didcot was to become the junction for a spur to Oxford. That valley, though, was old and wide, so that Brunel had no need to follow the river's sinuosities but could draw the great sweeps which still dominate the map. The Thames itself was crossed twice more in Berkshire, at Basildon and Moulsford: the bridges here are less striking than that at Maidenhead and lack its stuccoed detail, but they have perhaps suffered less from later widening works.

From Didcot on to another village junction near the crest of the line at Swindon (from which a branch, departing northwards on a long embankment, was to run through the difficult Cotswold country and a substantial tunnel to Gloucester), the fairly level and dry land presented no great difficulties for the main line, and a temporary terminus was set up at Challow in mid-1840. But meanwhile progress was being made at the Bristol end and Brunel, who lacked a school of trained assistants, found himself working twenty hours a day as he shuttled in his 'black hearse' *britzka* between various sites, London, and Bristol. The latter city was also the centre of that shipbuilding activity which was to carry across the Atlantic the name now assumed by the railway company—the proud name of 'Great Western'.

Here in the valley of the Avon there was little elbow-room. From Bristol to Bath the main river had to be crossed twice, while tributaries and artificial waterways called for yet further bridges. Some of these works—which included one bridge of attractive Gothic arches and another which afforded a good example of Brunel's use of timber—have vanished with the years; but five of the seven tunnels in the Twerton area remain as Brunel built them, the longest exceeding 1,000 yards and another exhibiting two fossils he had proudly excavated. Brunel's handsome viaducts at Twerton and at Bath station survive, too, and there are also heavy earthworks.

By 1841 the eastern section of the line had passed Chippenham by way of Wootton Bassett and a long curved embankment; and, as on the London and Birmingham, there remained only one tunnel to be opened before Brunel could write *finis coronat opus*. But, as not only Kilsby but Littleborough had proved, it was easy to underestimate the difficulties of a tunnel; and just as Littleborough had added a quarter of a mile to Kilsby's length, so Box was to add a quarter more to scale nearly two miles. It was also up to 300ft deep: it ran through mixed strata of clay and marl with only the few hundred yards at the eastern end of that 'great' oolite which could be left unlined: and in another respect it was unusual. For unlike all the long bores hitherto built on canals and railways, Box was not a true summit tunnel. Brunel's survey had the remarkable feature that for 70 of his 118 miles the ruling gradient was less than 1 in 1,000 (for the most part only 1 in 1,380), and for 40 more was increased to a mere 1 in 750. The 'steep' gradients were all contained between Swindon and Bath, with the sharpest section of all—three miles at 1 in 100—running through the tunnel itself.

This fact added to the fears concerning the Box tunnel expressed not only by the egregious popularist Dionysius Lardner but by a number of doctors of medicine too. A hilarious anthology remains to be compiled from the imbecilities uttered by those practitioners who, from pharaonic times to the age of BMA pamphlets, have persisted in stepping outside the narrow limits of their craft; and if this ever comes to be written one of its richest chapters will be that devoted to the prognostications of those Harley Street Hanoverians who did not appreciate the difference between acceleration and velocity but were as sure that tunnel travel *ought* to be bad for the body as their successors have been concerning smoking or high-altitude sports. The Box tunnel attracted more than its share of such comments.

Brunel, however, had more serious problems to face. From 1837 to 1840 a ton of gunpowder was fired each week, and a hundred horses were kept constantly at work hauling the 30 million bricks needed for the tunnel lining. Even with 4,000 men (of whom over 100 died) crowded on to the working faces, *finis* proved the hardest of words to write; and when the Box tunnel was completed in the summer of 1841 it was nine months behind schedule.

Now, though, the engineer had one of his moments of triumph; for even if Brunel did not deliberately align the tunnel so that the sun shone through it at dawn on his birthday, he confessed to a love of *la gloire* in its full and Gallic meaning. The shorter tunnels towards Bristol had been fairly plainly treated, but at the west portal of Box Brunel made his grand gesture. The style of the façade is still simple and dignified; but its proportions are vast, with the entry itself far higher than the bore behind it. This portal, though, had at least a psychological function; for it was visible from the turn-pike road which the railway here rejoined, and so nobly advertised the splendour and safety of the Great Western railway. It is sad that, the life of Bath stone being only a century or so, its details are rotting away in an age too poverty-stricken even to maintain what the past so splendidly built.

After such talk of glories, the name of Swindon sounds bathetic. Swindon is a railway town: therefore Swindon is a joke: so runs the English syllogism (and the French feel much the same about Laroche-Migennes). But in fact this junction below an old Wiltshire hilltop village holds a good deal of interest.

Brunel needed to select a point approximately half-way along his route where locomotives could be serviced or changed and where passengers too, in an age of four-hour journeys in trains without any amenities, could drop their waste and take on fresh supplies. Stephenson had created Wolverton in Buckinghamshire for this double purpose; but Wolverton was to be supplanted as its system grew by a site which was also a major junction, and apart from its carriage works became a mere refreshment stop—and not even that in the age of dining-cars and train toilets. Brunel and Daniel Gooch, with their vision of a planned Great Western network, could look further ahead and choose one of two strategic points to become their locomotive centre.

These points were Didcot and Swindon. They were about equi-distant from the geographical half-way mark, but Swindon recom-mended itself as being more nearly half-way in terms of engine work. Brunel seems to have thrown something onto the grass there to mark where he intended his station to arise, though it was probably *not* his top-hat. And southwards he laid out what may be Britain's best artificial township apart from Saltaire, with its orderly terraces of artisan's and foreman's cottages in local stone, its library and institute, and above all the church built for him by Gilbert Scott. It would be pleasant to think that this handsome piece of early Gothic revival was dedicated to St Mark as an act of filial piety by Isambard, though the Brunels (like most of their coeval engineers) were not religious men.

Some time before the first cup of *ersatz* coffee was forced down in a twenty-minute Swindon halt, however, Brunel was facing one of those setbacks which marked his life. For whatever the merits of the broad gauge itself, the 'baulk' track was not living up to expectations. There were complaints of uneven riding and even derailments: for these either the road or the rolling-stock might be responsible. But what could only be attributed to the track was the fact that normal settlement as well as washouts eased ballast away from below the timberwork and left the piles holding *up* what they were supposed to hold *down*. Like those who had favoured close-set stone blocks, Brunel had failed to appreciate that a certain resilience, rather than absolute rigidity, was desirable in a railway track.

In the first stage of public operation, with only a few miles open west of London, these track shortcomings became so serious that there was a movement to drop Brunel and the broad gauge together. Had this succeeded, one of Britain's greatest engineers might now be remembered primarily as the builder of the ships which were so far in advance of their times, and would be dismissed in a paragraph so far as railway history is concerned. But Brunel's personality and Gooch's ingenuity in keeping the traffic moving won through; and in any case the Great Western directors in London and Bristol had other problems to cope with. Foremost among these was the rivalry of the London and Southampton railway, which had been promoted a little earlier.

As we have seen, there was in the Napoleonic period great agitation for better communication between London and Portsmouth; but with the coming of peace the naval harbour had so diminished in relative importance that it could be served by a branch line to Gosport—a town today without a train service, perhaps in testimony to the fact that in Britain, as opposed to continental countries, it was only at the very dawn of the railway age that military considerations were thought important. Southampton by contrast, though in the 1830s somewhat depressed as a passenger port, fed a growing London with its much-needed fresh fish.

But with this shift of accent westward the proposed line was coming close to GWR country; and when there were suggestions that Bristol itself could be served by a branch from Woking the promoters of the direct route realised that they had a potential competitor. Already in the mid-1830s the railways believed that they had no more to fear from the canals as freight carriers than from the stage coaches for passenger transport: thus, the GWR used a canal to transport its materials before acquiring and then casually and illegally abandoning it. But the age when each line was a self-contained enterprise was passing, and as it did so that of outright rivalry drew in sight. The name the Southampton railway was to adopt, even before it was complete, as a tactful gesture to Portsmouth—the name, that is, of the London and South Western railway—implied an obvious threat to the trade of Brunel's line.

This book is not the place to set out once more the battles which developed between the two systems before either ran a train. It is enough to note that, as with the London and Birmingham, some form of joint arrangement with the Great Western was suggested but did not materialise. In the event, the Southampton line was to have its own London terminus south of the Thames in Vauxhall (which later became Nine Elms goods yard, and was eventually erased) and was built to the standard gauge.

The line was authorised in 1834, the first engineer appointed being Francis Giles who had worked on a ship canal project in the area. Giles, an ageing man by the standards of that time, was to do reasonably well on the Newcastle and Carlisle where he followed another man's survey; but he lacked the experience, interest or energy to cover eighty miles of virgin territory. He was asked to resign; and in 1836, when work was going ahead well on the Grand Junction, Joseph Locke took over. He immediately discharged a host of minor contractors but retained, of course, Thomas Brassey and his superbly organised navvy gangs. This team was later to work together in Normandy and inspire the comment (so ironic-sounding today) *Mon Dieu, les anglais, comme ils travaillent*, even though the 'anglais' between them spoke thirteen languages.

Meanwhile there was a might of muck to be shifted between London and the sea. Locke's predilection against tunnels of more than a few dozen yards, and his readiness to adopt something less than a direct route, showed clearly here, and the engineer was also prepared to relax his gradients to accept sixteen continuous miles at 1 in 250—a figure considered steep at the time. But the massif of downland chalk between Basingstoke and Winchester still had to be defeated. In defeating it Locke carried out what is believed to be the greatest of all the world's earth-moving programmes, the navvies disposing with their primitive aids of over three million cubic yards of spoil at a rate of about five million tons every year. Most of

SOME HISTORIC TUNNELS

15 *Top* Primrose Hill, London
16 *Bottom left* Box, Wilts.
17 *Bottom right* Shugborough, Staffs.

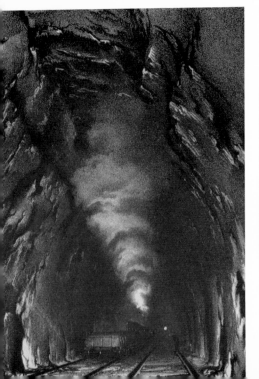

this was the Hampshire chalk which shoulders the 90ft-deep cuttings and embankments of Micheldever, but in addition the
Chatham ridge in Surrey posed its problems, and on the Gosport
branch there was some treacherous clay which almost bankrupted
Brassey.

Though there was a substantial north-country shareholding, the
Southampton line was more a 'London particular' than were the
GWR and L & B/GJ systems, where the initiative and capital had
come at least equally from the provincial terminals. The LSWR
was completed in 1840, and soon after it was opened to the public
with a naval salute of guns Londoners had the chance to compare
travel over three types of road. On the way to Birmingham there
was Stephenson's I-sectioned track partly laid on stone blocks; and
on the way to Bristol there was Brunel's unconventional system, still
suffering from teething troubles. On the way to the south-west,
however, there was a track improved from that which was already
proving its value on the Grand Junction, a road laid throughout
with doubled-headed rails supported by transverse timber sleepers
resting on ballast. It was this last which provided the smoothest
travel, just as (experience was to show) it called for the least
maintenance.

Because this 'parsons and prawns' route boasts no great works of
brick or masonry it lacks something of the historic appeal of its
compeers. Because Joseph Locke was neither a cautious pioneer like
Robert Stephenson nor a contentious innovator like Isambard
Kingdom Brunel he is even regarded as the least of the three, with
his name evoking such faint echoes in the public mind that he may
be more honoured in Normandy than in his homeland. But if this
neglect is a fact it is a regrettable one; for Locke, whatever he may
have lacked in personal glamour and however unostentatious his
works, showed surer judgment than either Brunel or Stephenson.
Nothing much that he did ever went wrong.

The last of the classic railways to be virtually completed by 1840
was the London and Brighton. Even in length it was the least of the
three mentioned in this chapter, and perhaps because it never
carried substantial freight and is today thought of as simply the
spine of a commuter network it is often overlooked that the Brighton
line had to contend not with the steady gradients of the Stockton and
Darlington, nor the comparatively level course of the Liverpool and

Manchester, nor the single summits of the Bristol and Southampton routes nor even the double watershed of the Birmingham
line, but with the three separate ranges of the Weald and the Downs
to north and south.

Furthermore, Brighton was only one of the line's twin objectives.
Beyond lay Shoreham, a port popular with William James—who on
this line too proved himself, if not the father of railways, at least
their godfather. This extension (which was useful in supplying
materials) was in fact opened before the parent line was completed,
as a token of the amount of freight traffic at first anticipated. But
the Prince Regent had so firmly launched Brighthelmstone in its
new role of London-on-Sea that even before 1840 the town's population was approaching 50,000; and so far had matters advanced since
the surprise success of the L&M that the function of this railway was
clearly to be to carry passengers.

It was to carry them, furthermore, for pleasure rather than
necessity. Once it had emerged that London was to be the node of
the nation's railway system (as it was of the roads, but *not* of the
canals), every long-distance company might want to have the magic
name of the capital in its title. But, even more than the LSWR, the
Brighton line was to be London's own.

After the opening of first the Canterbury and Whitstable and
then the Greenwich railway, developments in the Home Counties
had been slow. In 1836 a short line had been commenced from near
Fenchurch Street to Blackwall which represented the finest flower
of the cableway principle, with an elaborate slipping system being
employed to serve intermediate stations; inevitably this became a
railway backwater. In the next year another brief railway began
operating between Shoreditch and Brentwood, Essex; and in 1838 a
branch from the Greenwich line was opened through Norwood to
(the present West) Croydon. Engineered by William Cubitt, this last
employed longitudinally-supported rails of a curiously squat form
and made use of the bed of an abandoned canal, though it also involved some heavy cuttings through clay. The junction at Corbett's
Lane, which was now handling a local train every five minutes or
so, became the site of the world's first signal-box—a curious,
lighthouse-like structure.

Now at last came the plans for something which, if not a great
trunk railway, should be more than a local facility—the prolongation

of the London Bridge – Croydon line from near Norwood to the coast. Like most developments south of the Thames this used transverse sleepers; but it may be worth mentioning here that as late as 1844 it was proposed to use on a London-area branch William Prosser's guided system, which incorporated wooden rails (a system actually employed on a Parisian line for some years), and that even Cubitt believed these practicable.

No fewer than six routes were suggested for the Brighton line, which proved very expensive both in the parliamentary 'expenses' which were often a polite term for bribery (£3,000/mile) and in land acquisition (£8,000/mile). Each survey had its vigorous adherents, but they fell into the general strategies of an indirect but easily graded route taking advantage of the river gaps to the west and of a frontal onslaught by a beeline of slightly over fifty miles. Surprisingly, Robert Stephenson favoured a devious line; but the government decided in favour of a direct route which he considered impossible, just as his father had considered Locke's Southampton route wildly expensive.

This track was surveyed by the younger John Rennie. The engineer in charge derived even more directly from the tramway age, for he was the John Rastrick who had assisted with so many historic projects. And a triumvirate was completed by David Mocatta, architect.

To restrict the gradients to 1 in 264—a figure which, it was calculated, doubled the engine-power required on the level—three long tunnels through the successive chalk ridges were pierced without great trouble at Merstham (1,831 yards, plus some very substantial cuttings, at a site close to that of the old tramway), Balcombe (1,141 yards) and Clayton (2,259 yards). These, like many other such, were whitewashed within and kept permanently gaslit. There were also two shorter tunnels. The portals were in some cases given architectural treatment, that of Clayton (north) being unadulteratedly Gothic; but the glory of the line was undoubtedly the Balcombe viaduct striding on its thirty-seven brick and stone arches 100 ft above the valley of the Sussex Ouse.

With its perfect proportions and elegant Italianate pavilions this is, indeed, one of the splendours of English architecture; and it could be argued that the leading contribution of the south to the railway scene was the proof by David Mocatta, a pupil of John Soane, that railway structures could be not merely function-serving, not merely decent, not merely impressive, but consciously lovely. The viaduct was built to carry about a dozen lightweight trains a day; and though within a decade there were to be 1½-hr non-stop runs, these at first averaged only twenty mph. (On all such early lines, indeed, it is surprising that after so much trouble had been taken to build routes saving a few minutes they were so underexploited.) After 130 years Balcombe is being pounded many times an hour by heavy electric expresses travelling at over a mile a minute. The time must come when this, like many such structures, will need either heavy renewal or a complete rebuilding which might appear cheaper. It is to be hoped that, when it does come, the nation's sense of the past is more lively than it seems to be today.

By the end of 1841 the great majority of English towns with a population of over 25,000 were in sight of a railway, and a Londoner could travel by train to Shoreham, to Bristol (and beyond), to Southampton, and via Birmingham to Manchester. From there Liverpool could be reached on the one hand and Leeds on the other; while connections to other northern cities had been made as is described in the next chapter. Branches, extensions and new routes were opening in the Home Counties, Midlands, north and west, whilst passenger-carrying systems for the moment detached from the main network included the Stockton and Darlington, the Newcastle and Carlisle and a number of shorter new lines and rebuilt tramways. These brought the total route mileage of railways in Britain up to nearly 1,500, comparable to the peak lengths of both the canal and the tramway networks. But those latter had evolved over decades and centuries respectively, whereas most of the railway work had been carried out in the previous five years.

The map opposite shows the disposal of the more important of these lines; but it does not represent Britain's railway system as it appeared on any given day, for before one trunk line was completed the first local working had begun on its continuation or rival. After the spate of parliamentary Acts passed in the mid-1830s there had been a brief period of recession, hiatus and stocktaking; but the gestation period of several years between the sanctioning and the opening of a line smoothed out most such trade and psychological cycles and in fact new railways commenced business at a fairly steady average of some 200 miles a year throughout the decade

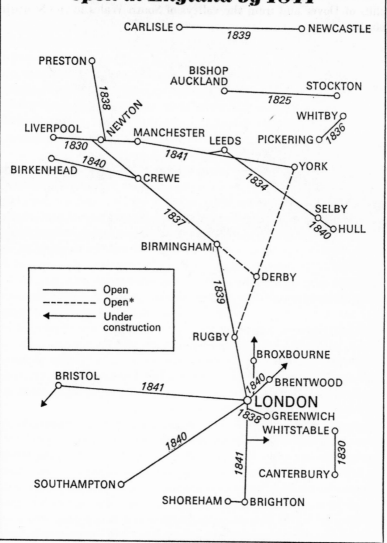

Figure 3
* for details see Figure 7

after 1836. And so, as the cannon rang out to salute inaugural trains in terminal after terminal, their reverberations were echoed by the shotblasts of navvies at work from the red cliffs of Devon to the white cliffs of Dover and from the valleys of South Wales to the Scottish border.

CHAPTER SIX

Midlands and North

The following two chapters carry the railway story on from about 1840 to 1846, the figures being approximate since building was continuous and it is necessary to select from a mass of work on a topographical as well as a chronological basis. In this period the national network expanded more than fourfold and had such wide-ranging sociological effects that there could be no doubt now as to the applicability of the phrase 'the Railway Age'. And through their success the railways attracted (as many technologies have in their second or third generations) a new type of man—not the engineering pioneer such as George Stephenson, not the godly early capitalist such as the Edward Pease who fought for righteousness throughout this phase, nor yet the grim and bearded patriarch who later came to dominate an assured industry, but one who in combining an eye to quick profits with a lack of social responsibility had nothing to learn from our contemporary property developers and whose alliances were as reliable as those formed in a cut-throat card game.

A general railway history would have to devote several pages to at least two such adventurers—George Hudson, the financier whose activities dominate the age and countryside of this chapter, and the Captain Mark Huish who had the morals but not the charm of Captains Foulenough and Grimes. (To illustrate the fact that a great industry had fallen into the hands of gangsters ready enough to break the law when they could not bend it, it is necessary only to cull from Huish's professional career the fragrant moment when he quietly forged the seal of a small and uncooperative rival.) But fortunately such commercial complexities are only marginally relevant to an account of railway building.

George Stephenson himself, for instance, does not seem to have been overmuch concerned with the power politics of his paymasters; and it was the Stephensons who continued to set their impress on much of that complex of lines which began in the Midlands at the close of the 1830s and which in the 1840s reached towards the Geordies' home country. It was they, for instance, who were responsible for the railway which first brought the market town of Derby into the railway network. This ran southwards to the L&B at Hampton-in-Arden, just outside Birmingham, by way of a fairly easy forty-mile route following the valleys of the Trent and Tame which had earlier been utilised by the canal system.

The line had been completed as early as 1839, and the next task in what was now clearly the manufacturing heart of England was to complete a rough quadrilateral by linking Derby directly to the Leeds area. This new and coalfield-serving North Midland railway passed through Yorkshire by way of Rotherham (for Sheffield) and Normanton, where it connected with the Manchester and Leeds railway: its seventy-two miles included some 200 bridges as well as seven fairly short tunnels. The crossing of the Derwent and its lateral canal was carried out with great ingenuity, but perhaps the most interesting feature remains the Clay Cross tunnel—which, unusually for the age and region, is adorned with elaborate medieval portals. In cutting this George Stephenson met the unexpected coal measures which furnished the first coal ever to be sent by rail to London. Even more unexpectedly, the sixty-year-old railway pioneer took an active interest in developing an iron industry in north Derbyshire and moved his home to a mansion outside Chesterfield.

Thereafter he tapered-off his railway activities, though working to realise two last dreams. The more important of these was the connection of his home colliery lines about Tyneside with the national network. This possibility had drawn in sight in 1840, when an extension of the North Midland to York (long a target for railway planners) had been opened simultaneously with the parent line, and it was achieved between 1841 and 1844 by way of the Great North of England railway. This straight and level line passed through Darlington—where a flat crossing with the Stockton line still epitomises the contemporary spirit of interconnection without co-operation—and bridged the Wear at Penshaw on Benjamin Green's fine Victoria viaduct of stone which is over 800ft long and includes four central arches up to 160ft long and 135ft high. The Tees was crossed by a bridge at Croft. An earlier interesting bridge in the Newcastle area had been built in 1838, the 'bowstring' iron span near Washington, Co: Durham; but for the moment the city of Durham was served only by branches.

The last of all George Stephenson's lines, and his favourite, was planned as a diagonal of the quadrilateral referred to above, and headed from Ambergate (between Derby and Sheffield) directly towards Manchester. After reaching the ducal estates about Rowsley, however, this project was to run into financial and engineering troubles; and it was not until the early 1860s that it eventually reached Manchester via 1 in 90 gradients, the fine Monsal viaduct, and that costly necklace of tunnels around Dove Holes which led enginemen to nickname this line 'the flute'. It also gave Ruskin the occasion for his famous 'vale of Tempe' diatribe, but scenic pleasure to a century of more broadminded travellers before its western end was closed and demetalled.

In 1848, after a final working visit to Spain, George Stephenson died. He had devoted his semi-retirement to the typically north-eastern hobby of growing monstrous vegetables; and by a pleasant coincidence his great rivals here were the Chatsworth horticulturalists, the Paxtons, so that amateur gardener and professional engineer met amateur engineer and professional gardener over the prize leeks. George Stephenson may or may not be entitled to be called the father of railways; but the first child of his invention was the miner's safety lamp and the last was a straight cucumber.

So Stephenson lived long enough to see through trains running from Euston to Newcastle. But the route through Birmingham totalled over 300 miles, and Hudson (who had financed its Great North of England section) had plans for a more direct link south.

Meanwhile his own network had been both augmented and challenged by a line between Rugby and the Derby area, serving Leicester (and, on a branch, Nottingham) and opened in 1840. This was the Midland Counties railway, which was eventually entrusted to that gentle ex-military-engineer who had moved even further away from the Stephensons than had Joseph Locke, Charles Vignoles. The engineering of the route is not of the first importance, its most curious feature being the asymmetrical Gothic portal of the short Red Hill tunnel built near a new junction called (to the disgust of Ruskin, who said there was no such place) Trent; what is of interest is that Vignoles suggested an alternative to Stephenson's simple rails *and* different from Locke's double-headed type. The first illustration here shows a selection of early forms and is taken from a contemporary source: the second illustrates modern versions of the two main sections, the introduction of the 'sole plates' which correspond to chairs being a fairly recent innovation.

The Vignoles rail—which was initially a little more expensive than the Locke type, but which proved both cheaper to lay and longer-lived once it was realised that it did not need to be set on longitudinal bearers—was to become the accepted form in America and on the Continent, and by a turn of the wheel of technical change was to return to its homeland as British Railways' standard well over a century later. But by 1840 Locke's double-headed form was becoming established in the midlands and north and beginning to

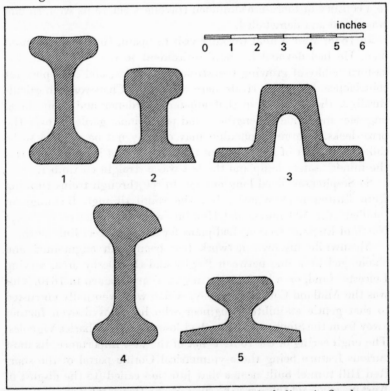

Figure 4 A selection of early rail cross-sections including 1. Stephenson, London & Birmingham Rly. 3. Brunel, Great Western Rly. 5. Croydon Rly, anticipating Vignoles

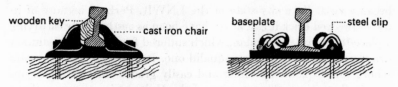

wooden key········   ········ cast iron chair   baseplate   ····· steel clip

Figure 5 Modern forms of bull-head (Locke) and flat-bottomed (Vignoles) types of rail, showing mounting

replace Stephenson's I-shaped bars there; and hence the Midland Counties eventually ruled against Vignole's rail and even instructed him to use stone sleeper-blocks.)

This should not be interpreted as suggesting that there was as yet any real degree of standardisation of rail sections. The Midland Counties line, for instance, was laid with rails weighing 78 lb/yd; but more than a decade later the Newcastle and Carlisle still ran over tracks of six different weights ranging between 42 and 82 lb/yd. Overall there was a continuing movement towards heavier sections and also towards longer rails, but this latter trend was controlled by the limitations of metal-working techniques and it is doubtful if a 60ft rail was rolled much before 1850. Three years before that date, Adams and Richardson had introduced the fishplate which tied together the ends of adjacent rails rather than leaving them to be secured in a common chair; after it there was to be little change in basic rail form for a century, though there was naturally a spreading use of the best practice.

Meanwhile the railway network was also expanding beyond the Pennines in the Manchester area. There the original GJ/L&M route by way of Warrington was an obvious candidate for a direct cut-off—the first of its kind—crossing the easy Cheshire plain from somewhere north of Stafford. Since a corresponding (and equally straightforward) line westward through Chester to Birkenhead had been opened in 1840, it was natural to choose the same junction. This was Crewe, laid out by Joseph Locke and built between 1840 and 1843 complete with town hall, schools, baths, gasworks, the Anglican Christ Church, and grant-assisted places of worship for Roman Catholics and nonconformists too. Curiously enough for an engineers' town, its weakest feature was its water and sewage arrangements.

In the second half of the nineteenth century—and after—Crewe

became virtually a city-state of the LNWR. Perhaps because of its
use of raw red brick it can never have been as aesthetically satisfying
a place to live in as Swindon, which suffered from the same sanitary
drawbacks. But it was not a squalid one. Its *raison d'être*, the point
where three splendidly direct and easily graded lines diverge sym-
metrically towards the extremes of the United Kingdom, remains a
monument to the assured trackwork planning of the early 1840s.

The Manchester cut-off itself was completed after some commercial infighting in 1841, and has one notable engineering work at
its entry to that city. This is G. W. Buck's 600-yard-long brick
viaduct which marches above the slate roofs and cobbled streets of
Stockport on twenty-two arches up to 110ft high. Costly as the work
was, it proved cheaper (as the south London companies had dis-
covered) to build viaducts rather than earthworks in the hearts of
cities where land costs were high—and, as the illustration shows, the
taller the structure the greater the saving became. The Stockport
viaduct carries today's heavy and high-speed traffic virtually unchanged except for widening.

One other line remains to be mentioned in this somewhat arbitrary chapter whose contents are summarised in the map opposite. It
represents a southerly crossing of the Pennine spine which succeeded

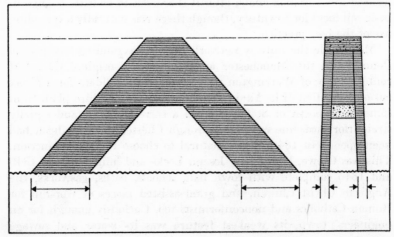

Figure 6 Additional materials (shaded) and land (arrowed) needed to
double the height of an embankment (left) and a viaduct (right)

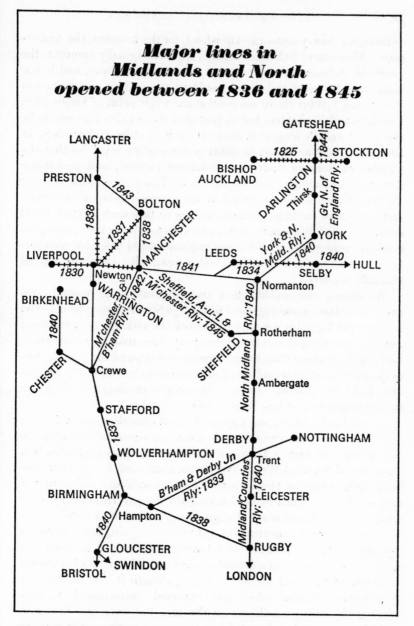

Major lines in Midlands and North opened between 1836 and 1845

Figure 7 Only the railways named are dealt with in Chapter Six. Lines shown ticked were open before 1835

—though at heavy cost—close to where, for the moment, the Amber-gate – Manchester link was halted. This was originally known as the Sheffield, Ashton-under-Lyme and Manchester railway, and it was clear that the great task on it would be the cutting of a tunnel at Woodhead, 1,500ft above sea level at the triple point of Derbyshire, Cheshire and Yorkshire. For at just over three miles this was to be easily the world's longest tunnel yet built to railway clearances. In addition it ran as much as 600ft below surface level (so that the number of working shafts was restricted to five), and penetrated five strata including a treacherous shale. The greatest difficulties of all, though, were those imposed by the extreme isolation of the site and its lack of facilities. Miners willing to face such hazards could demand up to £20 a week—a fantastic sum for labour then, though it is worth remembering that a top engineer could price his own ser-vices at the equivalent of £50,000 a year, plus expenses and all virtually tax free.

Woodhead, the world's first true mountain rail tunnel, was another of those works regarded by George Stephenson as impossible; but in 1838 Vignoles not only undertook the task but underwrote it heavily by buying shares in the company. Less than two years later, though, he realised that he was committed beyond his means. And despite the loyal intervention of the company chairman, the ubiqui-tous Lord Wharncliffe, he was discharged in disgrace when the tunnel began to run into its inevitable setbacks.

Joseph Locke, that great opponent of all tunnels, then took over, his first reformation being the very Lockeian one of rationalising the labour force of over a thousand and appointing as contractor the experienced Nicholas Wood. Work then continued for six years un-interrupted either by bleak winters or by the accidents which injured nearly 700 men, until by Christmas 1845 the government inspector could pass 'the finest work of engineering he had ever seen'.

Soon afterwards construction had to begin again, for to bring in revenue as soon as possible the company had decided to build two single bores rather than a double-tracked one. The parallel operation naturally went ahead somewhat faster, though it was interrupted by a cholera outbreak which was contained—miraculously so, con-sidering the living conditions on the site. But even the first bore could be accounted as Joseph Locke's greatest work, and it perhaps atones for the fact that this fine engineer, never lucky with his bridges, had meanwhile run into trouble nearer Manchester. For there the timber viaducts at Dinting Vale and Broadbottom were showing signs of subsidence and demanding to be replaced by iron spans which themselves were to prove unstable, while a masonry viaduct at Stalybridge had collapsed with the loss of eighteen lives thanks to a contractor's Romanesque habit of building his piers of unmortared rubble encased in a thin dressing of stone.

For a number of years Woodhead remained the world's longest railway tunnel through rock. But—partly due to tight clearances and to the operating oddities caused by the twin tunnels being built on a 1 in 200 gradient—it did not wear well; and when after the second world war a local electrification scheme would in any case have necessitated rebuilding it was decided to cut a completely new tunnel (and this time a double-tracked one) some thirty yards to the south.

Today the gargoyles over the western portals of the original bores are left to erode away unshaken by traffic, though one tunnel has found a new use as a duct for power cables. But to see the achievement of Vignoles, Locke, Nicholas Wood and even Lord Wharncliffe in perspective, it is worth remembering that the tunnellers of the 1950s worked with mechanical drills instead of hammers and chisels and with custom-compounded high explosives instead of 150,000 tons of black powder, that in real terms their budget of over £4 million was about twice that of their predecessors, and that they also had the advantage of reference to the century-old surveys.

Yet these later engineers were able to economise by only a few hundred on the original labour force, and to improve on the six-year schedule of their forerunners by little more than one year.

CHAPTER SEVEN

To the South and West

While these developments had been taking place north of Birmingham, the railway builders had been almost equally active in the south. The oldest route not yet discussed was indeed nearly a contemporary of the Brighton line, and it is at first sight surprising that the South-Eastern railway was not built still earlier considering Dover's long-established precedence as a continental bridgehead. Following a scheme of Telford's, the Greenwich railway had in fact headed in the general direction of the Dover road; but it soon became clear that the contours along this route (and not least on the final drop into the port itself) could not be accommodated to the easy gradients of contemporary railway engineering.

The survey finally chosen was hence an unusually indirect one. At the Channel end it approached Dover by way of Folkestone, so compressing all the heavy building work into its final seven miles; and at the London end (following that parliamentary agitation for the 'concentration' of the capital's rail traffic which had failed to group the western lines into one terminal) it utilised the existing tracks running southwards towards Brighton from London Bridge, not turning eastward at a right angle until the Surrey junction which was later to be known as Redhill and to develop into an important railway town.

For the most part the new line thus ran across the very easy country of the Vale of Kent. After a tunnel at Bletchingley (1,300 yards) it passed through Tonbridge and Ashford, where the SER set up a planned town which is claimed to have the only back-to-back housing in the south; and only the Medway viaducts and a few light earthworks were needed to keep the gradients throughout down to short stretches at 1 in 220. The line also ran almost straight for nearly fifty miles—a feature of British railway engineering unchallenged even in Fenland, though the longest *dead* straight runs for eighteen miles south of Hull.

THE TIMBER BRIDGE TRADITION   18 *Top* Tregeagle, Cornwall
19, 20 *Middle and Bottom* Wallsend, Nthbd.

EARLY LONDON TERMINI   21 *Top* Euston
                        22 *Bottom* Blackwall

This routing implied that to leave London the SER had to work over the tracks of three other companies, branching off from first the Greenwich, then the Croydon, and finally the Brighton railway's metals. In slightly later years, when the L&G became amalgamated with the SER and the L&C with the L&B and when track occupation over these twenty miles was increasingly dense, this was to give rise to difficulties; but when work began in 1837, such inter-company rivalries were only a shadowy threat. It is perhaps note-worthy that the surveyor of the SER was the William Cubitt (in himself an anomalous figure, since he was a second generation con-tractor rather than a pure railway engineer) who had built the Croydon line but had not even tendered for the Brighton one.

Cubitt used a rail weighing over 70 lb/yd and closer to the bull-head form than to the inverted-T shape which had been employed on the Brighton line. This he mounted on sleepers which were triangular rather than rectangular in cross-section, for the question of roadbed formation was still a very open one. In this context, the comment by a Board of Trade inspector which Dendy Marshall quotes in his history of the Southern railway is of interest:

I am inclined to believe that this plan [of using triangular sleepers] will prove more economical than any other, and superior to all in smoothness of motion, except perhaps the rails laid on longitudinal sleepers. I cannot help remarking on this subject that it appears to me to be a great improvement that railway companies are inde-pendent in their arrangements, so that as no two of their engineers think alike as to these details, new plans are continually tried. . . . Thus, for example, fish-bellied rails, which were at first conceived the most perfect, have been entirely given up, and the stone blocks of the first railways seem also to be becoming obsolete. The travelling on Mr Cubitt's rail is certainly very easy.

The line itself helped to bring much-needed agricultural produce into an expanding capital, but the SER was overwhelmingly a passenger railway and drew some 90 per cent of its early revenue from this source.

A terminus outside Folkestone was reached in 1843. The immense effect of the coming of the railways on Britain's harbours generally is described in a later book in this series, but it should perhaps be noted here that, although the two Kentish ports seem nowadays as in-separable as Calais and Boulogne, Folkestone was in 1840 only a fishing village whereas Dover had at least 2,000 years of supremacy behind it. The railway came to *serve* Dover, and as a by-product virtually *created* Folkestone.

Behind this latter port lay the Foord gap which William Cubitt crossed on his most famous work, a 100ft-high viaduct of nineteen arches. But beyond awaited the approach to Dover itself. Early projectors had hoped to reach this by way of a track built out over the foreshore; and in fact the final stretch was laid on timber trestles, spanning tidal waters, which endured for eighty-five years. But most of the track had to be carved out of the chalk cliffs, and these six miles were perhaps tougher than any hitherto faced by a railway.

The three tunnels alone—which, from the Folkestone end, were Martello (530 yards), Abbotscliff (1,940 yards) and Shakespeare (1,390 yards)—totalled well over two miles and had unusual features including the fact that one was built with twin bores terminating in high and pointed arches; horizontal working shafts extending out to the cliffs were used as well as those conventional vertical ones whose brick drums still mark the course of the line. Between the first two tunnels, too, the railway had first to cross the two miles of a landslide-prone area of gault and greensand known as the Warren, and then be carried along the first sea wall built for railway purposes. Also two miles in length, this was some thirty feet thick and sixty high.

From the public viewpoint, though, the great achievement was the removal of an entire hillock—Round Cliff Down, which no longer features on a map—to make way for the South Eastern railway. The army was called in, led by a lieutenant who had blown up the wreck of the *Royal George*; and in January 1843 more than eight tons of gunpowder were electrically fused together in the working bores and then detonated, with the assembled populace watching from pavilions a few hundred yards away in quaking delight as a mass of chalk 300 feet long and seventy high surged out towards France. This may not have been the biggest deliberate explosion in civil history, but it was certainly the biggest for which sponsors sold ringside seats.

So successful were all these works that Cubitt reached Dover in February 1844, not only on time but well within his budget. The company then pressed on to the building of extension lines—in the same year to Maidstone, and in 1846 via Canterbury to Margate. But at the same time its neighbour, the Brighton line, was also colonising this southern area of ports, of seaside watering places capitalising on the new and more widely spread wealth generated by the industrial revolution, and of inland agriculture.

London-based though the Brighton company was, its major early extensions took place at the maritime end of the line: they are in fact often referred to as the 'east' and 'west' coast projects, to the confusion of those who first think of these phrases in terms of connections to Scotland. The former is the earlier, and produced one civil engineering work of the great pioneering form as the line backed out of Brighton across the dry valley of the London road.

There Rastrick and Mocatta built a curving viaduct 330 yards long and comprising thirty-seven arches. The central of these, 67 feet above the valley floor, was widened to accommodate the posting-road whose 4½-hour service to the capital had so recently been put out of business, and as a result the whole composition achieved a Roman grandeur. The quality of the workmanship was proved when, almost exactly a century later, a German bomb dropped at random freakishly richochetted onto the central arch: this fell, but its neighbours stood fast and the structure was swiftly repaired.

This 'east coast' line could never have hoped for heavy traffic since its proximate terminus was only the county town of Lewes. Perhaps for that reason Rastrick allowed himself to climb out of Brighton over gradients steeper than 1 in 88—a figure which would have been intolerable to the pioneers but was to become increasingly acceptable, at least in branch line use, after 1845. But the rest of the route could use the wide coastal plain, by way of which it not only reached Lewes in 1846 but immediately continued, via the new continental port of Newhaven, towards Eastbourne (another new creation) and the ancient town of Hastings. Earlier, though, a 'cut-off' route had been built to Lewes from the main line south of Haywards Heath.

In the westerly direction, Worthing was reached in 1845, Chichester in 1846 and Portsmouth in 1847. The only engineering works needed here were bridges across tidal creeks, whether on the main line or on early branches such as that to Hayling Island, and some interesting examples included a telescopic span across the Arun. (This was not the first railway opening bridge, though, since the Greenwich railway already had one.) Portsmouth itself was in South Western country so that a joint-running agreement had to be entered into; but the LSWR was slow in improving its communications to the naval centre and in the 1840s appeared more interested in schemes involving Salisbury. Although this cathedral city had only an indirect link to London it was indeed being prematurely hailed as the 'Manchester of the south'.

Mercifully this change never came about, but the phrase is worth quoting since at this period the south seemed envious of northern industries. Thus, Newhaven was supposed to become a south-country Liverpool. For major progress south of that classic line from the Severn to the Wash, however, it is necessary to turn to what Brunel was doing in the west country.

Having achieved their link to London, the Bristolians were pressing ahead with a complementary scheme to Exeter and beyond. The Bristol and Exeter railway itself was formally an independent line with its own terminus (but a physical connection to the GWR) at Bristol; but for its first few years it was operated by, and it was always closely linked to, the earlier railway. Hence it inherited the broad gauge and Brunel, carrying the latter on towards the total of 1,100 miles of track for which he was to be responsible.

The first section was easy to build, and a connection to the then-important port of Bridgwater had in fact been completed by the time the inaugural train from London reached Bristol in 1841. Just beyond, where the line crossed the river Parrett, Brunel attempted to outdo his work at Maidenhead with a still flatter brick arch; but in this case the critics' fears proved justified, the abutments began to yield, and Brunel had to leave his centering in position until he could replace the masonry bridge with a timber one. The whole operation was conducted with the utmost discretion, for Brunel's negative public relations could be as effective as his positive ones.

Other landmarks of this part of the route are the two-mile Ashburton cutting and another cutting at Uphill near the junction for Weston-super-Mare. (Three other short branches were also included in the scheme.) This latter was crossed by a remarkably graceful road bridge which remains in use. Elsewhere the railway ran straight for many miles across the lonely Parrett marches, so that Taunton was reached in the following year.

But beyond the Somerset town the going was very much tougher across the col between the Black Down hills and the edge of Exmoor; and on his way to the thickly-lined, 1,000-yard Whiteball tunnel

Brunel had to stiffen his gradients up to a short run at nearly 1 in 80. On the further side there was a much more gentle descent to Exeter, which was reached in the summer of 1844.

In the mid-1840s the deep west was, like all agricultural areas, in a depressed state; but either in spite or because of this economic fact it was resolved to press on immediately to the civil and naval ports at Plymouth. This new enterprise, the South Devon railway, was sponsored jointly by the GWR, the B&E, and another Bristol company. And if the South Devon was to be the scene of Brunel's greatest mistake, it at least began with the notable work of building what must be the most famous length of track in Britain, the necklace of seven short tunnels interspersed with lengths of sea wall which runs between Teignmouth and Dawlish. Its construction was hindered by the winter gales of 1844/5, it has always been prone to assaults from landslides on the one hand and breakers on the other, and only very recently have new methods of ballast consolidation secured its roadbed against the lash of the ocean. For here, at '. . . Brunel's wall/Between the red cliffs and the rise and fall/ Of Channel tides . . .', the railway presses closer to the sea than along the Kentish seaboard or even in Liguria.

In 1846 steam trains were running to Newton Abbot and construction was advancing towards Totnes. But something much more surprising was happening back at Exeter, where between the rails Brunel was now laying lengths of curiously slotted iron tube.

This was part of what Devonians were to call the 'atmospheric caper'; and though atmospheric traction is dealt with in the companion to this book the subject calls for some mention here since it affected not only railway architecture but railway civil engineering. Briefly, then, out of all the rivals to the steam locomotive which had been experimented with during the first decades of the century there was one which had not merely not been exploded but which— thanks to the efforts of a gasman named Clegg—came in for a revival in popularity after 1842. This scheme (which embodied the basically sound ideas of separating rolling stock from its heavy prime mover and of load-equalisation on a busy line) had evolved from the original idea of blowing whole trains through tubes as bills are blown towards the cash desk of a departmental store, and now took the form of hauling carriages behind a piston sliding in a partially evacuated pipe.

Some test lengths proved spectacularly efficient, particularly in hill-climbing, and in 1845 the railway world was divided between those who believed that the future lay with atmospheric traction and those who felt that certain mechanical difficulties were not accidental but intrinsic. This division was not confined to England alone; for whereas in the suburbs of Paris an atmospheric railway was opened which ran for over a decade, and another proved fairly successful in Ireland, the system was considered for a mountain line in Austria but rejected after its engineer had seen conventional locomotives coping with long 1 in 75 grades in America. In Britain Robert Stephenson regarded the 'caper' with the scorn one would expect; but Cubitt and the proprietors of the Croydon railway took a more optimistic view.

It is indeed worth leaving the Devonshire coast for a moment to see what was happening in the London suburbs. There, one of a number of new lines proposed south of the Thames was to continue from West Croydon to fashionable Epsom. Atmospheric traction was considered a possibility for this; and in 1846, as a first step, the existing line from Norwood to Croydon was 'atmosphericised'—the same busy section as was to be the scene of one of Britain's earliest essays in electrification in the next century. The experiment is relevant to this book in two ways—because the difficulties of constructing junctions involving atmospheric pipes implied the building of the first flyover tracks, and because the Croydon pumping station was later transported brick by brick to the waterworks in Crown Hill where it still stands. The more elaborately Gothic passenger stations have, alas, now vanished.

Brunel studied the Croydon trains and was impressed: the atmospheric system with its promise of high speeds up steep gradients was one to appeal to a man whose vision outran his concern with what he dismissed as minor mechanical details and who may have resented the restraints which the capabilities of the steam locomotive put upon his dreams. Many years before he and his father had indeed played with a high-pressure 'gaz' engine which was intended to outmode steam. Now, he decided, the South Devon railway should be an atmospheric one.

The only section ever operated by pipework was that between Exeter and Newton Abbot—which, ironically, was almost level. Along this pumping stations were erected at intervals of about three miles. They were built in a pure Italian style, with the boiler

smokestacks being treated as campanili, and showed Brunel's architectural gifts at their finest and reaching a balance between fitness for
purpose and an appreciation of a Mediterranean land-and-sea-scape
apparently unrealisable in our present age. Happily one of these
gracious buildings survives at Starcross, where it is not unfittingly
used as a chapel.

The atmospheric caper itself, however, did *not* survive. After little
more than a year, with the Croydon scheme also showing signs of
failure, Brunel recognised that the difficulties of shunting and
vacuum-maintenance were insuperable, and with a certain courage
threw in his hand. Meanwhile, though, the damage had been done,
for the line from Newton Abbot on through Totnes to Plymouth
(with a branch to Torbay), which was completed in 1848, had been
laid out on the assumption that the atmospheric system would do all
that was expected of it.

In the first instance Brunel had chosen a more direct, hilly and
inland route round the edge of Dartmoor than he would have done if
planning a conventional railway. He had compounded his error by
engineering the line as if it were a tramway incorporating cable-
hauled inclines, and so far from averaging-out his gradients he had
deliberately grouped them into four 'planes' at the valley-crossings of
which the steepest was at the almost unprecedented figure of 1 in 37.
The all-conquering atmospheric system was supposed to cope with
these by the use of extra-large pipes. Finally (and with the least
excuse) he had assumed that the speeds attainable with atmospheric
traction would be such that only a single line would be needed, and
had not taken into account the fact that such time as might be gained
en route would be lost in manoeuvring at stations. Such was Brunel's
faith in his atmospheric caper that even the Exeter–Newton Abbot
section had been built, contrary to all contemporary main-line
practice, as a single track; and this combination of steep gradients
and singling was to hinder movement down the Devon coast long
into the century.

Two other points concerning the South Devon railway are of note.
One is that the short Marley tunnel appears to be among the few
where the legend that it was imposed by a landowner jealous of his
views can be substantiated. This tale is told of more railway tunnels
than there are canal tunnels with ghosts, but usually the reason for
what today appears an unnecessary bore is simply that for many

years it remained easier to mine than to excavate a cutting. The other interesting point is that the South Devon made considerable use of timber viaducts. But these, though found wherever Brunel worked in the deep west, are more characteristic of the final extension of this great route which belongs to a later chapter.

Meanwhile the Great Western empire was extending northwards too, though here the company wore hats bearing various local slogans and went arm-in-arm with some improbable co-belligerents as it entered an area of company wars. The first such link between the Midlands and west took the obvious form of completing, via Gloucester and Worcester, the triangle sketched out by the routes from the capital to Bristol and Birmingham.

This link was completed in 1847, though the northern section had been opened four years earlier. From an engineering (or, at least, operating) point of view, its most interesting feature is that the descent from the plateau of the Black Country to the plain of the Severn, which was paralleled by a long flight of canal locks at Tardebigge, was achieved by a two-mile length at 1 in 37. This run, the Lickey Bank, was to be a steepness feature of Britain's railways until the end of the age of steam, and demanded additional locomotive power. But that it could be included at all in a conventional main line shows how early the mechanical engineer was coming to the aid of his civil colleague and allowing the latter to relax his earlier demanding standards.

The Bristol–Birmingham route consisted of separate lines meeting between Gloucester and Cheltenham, Brunel being responsible for the southern section through the short Wickwar tunnel and William Moorsom for the longer, and exclusively standard-gauge, northern part. Two other links between the west and Midlands, though owing much to northern finance, were built entirely by the GWR. That company had now reached Oxford and saw two routes onwards—via Banbury and Warwick (as surveyed by Rennie long before) to Birmingham, and via Worcester, where the new line would cross that from Bristol to Birmingham, to Wolverhampton. All this was new country, though the completed routes would compete with the Euston ones. But the Great Western subsidiaries also planned their own link between Birmingham and Wolverhampton and a continuation of it by way of Shrewsbury and Chester to the Mersey.

Considerable though the total distance of over 250 miles was, few heavy engineering works were needed: exceptions were the section near Ruabon in Denbighshire where (beside Telford's canal) a main line entered Wales for the first time, some miles in the Black Country itself, and perhaps the crossing of the valley of the Shakes-speare Avon at Warwick. Though work was to be interrupted, both lines were started in that frenzied period of the mid-1840s when— it has been variously stated—a quarter of a million men were in-volved directly in railway construction and three-quarters of a million were making materials for new lines, in addition to the 50,000 or so involved in actual operation. (Another way of stressing the activity of the period is by pointing out that between 1830 and 1860 some 30,000 bridges were built, more than had existed in the kingdom before). The active building time amounted to only a few years, and the whole scheme is a striking illustration of the way in which, after less than a decade, such enormous works had become almost routine.

But from the end of the broad gauge at Wolverhampton up to the Mersey the GWR was directly invading the territory of the northern companies and the political battles were vast and complex, extending over a full decade. They even led to physical bloodshed comparable with that which developed at Mickleton, Worcestershire, when 'General Brunel' led his troops of navvies out to capture that singu-larly useless object, a half-completed tunnel, from an unsatisfactory contractor. But they are outside the scope of this book.

Not all the purely western extensions have even yet been men-tioned, for in 1844 Brunel had begun a long line from Gloucester beyond the Severn, Wye and Usk and so along the coast of South Wales to Newport, Cardiff, Swansea and the Pembroke headland. This line, which despite its extent of 150 or so miles demanded few important works except for further timber viaducts and an iron bridge at Chepstow which is described later, was naturally built on the broad gauge: it also employed a new form of rail devised by W. H. Barlow which aimed at dispensing with the longitudinal members of the 'baulk' track by widening the flanges of the rail itself so that they bore directly on the ballast. Barlow rail was also experimented with by some standard-gauge companies, but proved shortlived. Fragments of it are reported to remain visible at Frome, Somerset, and to be ignominiously incorporated into a bridge over a disused canal outside Swansea.

In parenthesis, it may be mentioned here that at the same time

W. H. Barlow's brother Peter was reviving in Kent the use of iron sleeper-blocks held together by tie-rods. This scheme proved even more abortive, though, and perhaps a more interesting reflection is that the father of both Barlows was also a railway engineer of some note. The number of family teams who worked on Britain's railways in the nineteenth century is indeed striking, for there are nearly a dozen instances to give the lie to the legend that talent usually skips a generation.

Reverting to the west another line in south Wales, the Taff Vale railway, ran inland from Cardiff to the Rhondda as early as 1840, largely following the old valley tramways. In this broken country Brunel had used the standard gauge on the principle that the Taff Vale was essentially a mineral line where the high speeds which were claimed to be a feature of his broad tracks would not be needed. It is clear from this alone that Brunel always underestimated the importance of that interconnection and possibility of through-running which wiser men had stressed at the beginning of the century.

Perhaps he cannot fairly be over-blamed for this, since parliament itself had long been ready to watch the railways of Britain grow up in piecemeal fashion and connect almost by accident. But after the 4ft 8½in and the 7ft 0¼in tracks had met at Gloucester, the distress to the travelling public and the chaos of goods transhipment (the latter being deliberately exaggerated by the porters of the standard-gauge interests when they knew that official investigators were watching) brought matters to a head. The Government turned aside from its Corn Law anxieties, and in 1845 a parliamentary commission was appointed to look into the gauge question.

From then on matters developed commendably fast by modern standards. A series of railway 'races' were officially sponsored and showed a real advantage to the western system in terms of speed; but against this the rest of England could point out that the length of standard-gauge track already laid was some four times the broad-gauge total. (And it *did* point out, with the world's first public relations officer being hired to 'create a public opinion' in favour of the standard gauge.) In 1846 the commission made what was perhaps the wisest recommendation possible at a time when W. E. Gladstone's attempt to regularise railway matters generally had just had a rough passage through Parliament.

By this new covenant, the GWR and its associates were to be allowed to expand on the broad gauge in the country which they had already made their own—which, roughly speaking, would be swept out by an arm extending from London to Exeter and then rotating clockwise across the deep west and Severnside to the heads of the valleys of South Wales. But they could not resist the penetration of even this area by standard-gauge companies, and they were themselves forbidden to press further 'unmixed' seven-foot tracks into the Midlands or Wessex.

Many such border-country lines had in fact been laid with a third rail to allow for mixed running, but now the broad gauge became so junior a partner that Brunel resigned in disgust as engineer of the Oxford, Worcester and Wolverhampton company and the GWR extensions northward became matters of commercial rather than engineering expansion. Soon standard-gauge trains were running into Paddington itself. The broad gauge was to serve the west for nearly another half-century, though even there there were invasions by the LSWR all the way from Wiltshire to Cornwall; but its unchallenged hegemony was so short that this may be the place to summarise its virtues and vices.

Leaving aside the fact that Stephenson's gauge got in first and grew so fast that it was bound to become accepted as the national standard with Brunel appearing the anecdotal 'only soldier in step'— and leaving aside, too, the qualities of the 'baulk' system of track-laying and of 'bridge' rails—would it have been better if Britain and the world had accepted (except in really mountainous areas where even the standard gauge sat unhappily) a gauge nearer to Brunel's seven-foot-plus than to Stephenson's five-foot-minus? The answer is, almost certainly, 'Yes', considering the comparatively low price of land and labour in the early railway-building age. For a small increase in wayleave costs and a somewhat larger one in the price of making engineering works, today's railways would be able to offer greater passenger comfort and more economical freight rates too. Even Britain's canals, it has often been pointed out, would have been in a stronger position had they been built to a 14ft (rather than 7ft) minimum.

Yet it cannot be claimed that Brunel ever took full advantage of that monstrous gauge which biblically-minded critics said was the broad way which led to destruction. His comparative lack of interest

in the steam locomotive—and his apparent willingness to see roads
built for mile-a-minute speeds worked in their first years at averages
as low as 15 mph—has already been mentioned. But he made
another error just as serious.

For at least as important as track gauge was loading gauge, the
all-round clearances imposed on rolling-stock by bridge heights,
platform widths and the like: it was this which set the ultimate limits
on both locomotive power and the carrying capacity of trains. Here,
as so often, Britain suffered and still suffers for being an industrial
pioneer: a comparatively densely built-up country of high land values
even when railways began, it fitted its trains more tightly to their
surroundings than did the other nations of Europe—let alone the
rest of the world. (In the United States, for instance, tunnels were
built nearly 50 per cent higher than in Britain.) In every age of railway history, up to the present one of overhead electrification, these
restricted clearances have posed a challenge to engineers; but Brunel
was so blind to the problem that he left even the space between the
two roads of his doubled tracks as the traditional 'six-foot way'. In
general, having achieved his broad gauge he did not seem to know
what to do with it: even his coaches were far less roomy than they
could have been, and the finest of broad-gauge trains had about as
much visual appeal as a flat-worm.

A version of the 4ft 8½in railway scaled-up all round to (say)
5ft 6in could have justified itself far more than the 7ft breadth ever
did; and after so much talk of two gauges it should be remembered
that many of the world's railways did in fact attempt compromises.
Perhaps the most compromising country of all was Ireland, which
worked out a form of weighted average of the existing English main-
line gauges before deciding that its national standard should become
5ft 3in, but later heretics were the Iberian countries at 5ft 6in and
Russia at 6ft: the latter in time reduced to 5ft, but Spain and Portu-
gal still retain their original gauge. Certain lines in the Americas
and (then) British empire also took to unconventional widths, though
Scotland had early been saved from adopting a nationalistic 5ft 6in.
By contrast, mountainous countries such as South Africa built up
successful networks of metric or 3ft 6in tracks.

At home there was only one other dissenting area; and to deal
with it this chapter must end with a sweep from the hills of the deep
west to the marshlands of the east as well as backwards in time. For

when in the mid-1830s the Eastern Counties railway had set out from Shoreditch with its eyes set on such far goals as Yarmouth its feet had been planted on a 5ft gauge track imposed by the less competent of its two engineers. This was John Braithwaite, who is otherwise remembered mainly as the co-designer of an unsuccessful 'Rainhill' engine: he now insisted that a decent locomotive must be at least 4ft 11½in wide, and managed to win not only his colleague Vignoles but also his directors over to this curious theory.

With little to draw on but shrunken ports, small market towns and underpopulated agricultural land, the Eastern Counties seemed to its critics to be an enterprise whose eccentric gauge (which had otherwise been used only on the Blackwall railway) hardly mattered since it was so remote from a swelling manufacturing economy. Certainly the line at first made painfully slow progress, despite the fact that a steam excavator seems to have been experimented with. Once clear of the Lea marshes, however, advance beside the level Roman road was somewhat swifter, and Colchester was reached in 1843 by way of a viaduct through Chelmsford. There the line languished for some years until it was extended to Ipswich by Joseph Locke and Peter Bruff.

In 1843, too, an independent railway was opened from Yarmouth to a station in Norwich which had been converted from a park rotunda. This single-tracked line had been conceived by the Stephensons as a stage in a great lateral trunk route, another part of which (again opened in 1843) took the form of a long and lightly-engineered branch from Blisworth on the London and Birmingham to Peterborough, establishing that cathedral city as a future railway *entrepot*. As a through route, though, the scheme failed.

Meanwhile yet another company—the Northern and Eastern railway, which was to some extent a rival of the Eastern Counties but was so closely tied to it that it too used the 5ft gauge—had crept out to Broxbourne, Hertfordshire; the engineer here was James Walker, who had been responsible for the Leeds and Selby line. The N&E was envisaged as part of a scheme, which went back through the Rennies to William James, to link London directly to Yorkshire; but it took from 1837 to 1845 to reach even Cambridge. From then on the colonisation of East Anglia proceeded more swiftly, the first route from London to Norwich being completed via Ely. But in 1845 the 5ft gauge had been narrowed to standard width, having endured well under ten years and been laid over less than ninety miles.

The invigoration of these lines, which were a decade later to become the Great Eastern railway, was largely due to George Hudson's shady enterprise, though it is more a tribute to the complete dependence of mid-century Britain on its railways that the system should shortly go on to construct such impressive monuments as the Chappel viaduct in Essex—a thousand-yard range of splendidly classic arches built up from seven million bricks to serve a sleepy rural branch. But in the age when they were laying out the skeleton of their network the East Anglian companies had such slender reserves that they must have been grateful that, through the nature of a country whose main hazard was the glutinous peat of the Fens, they were rarely called on to build anything more ambitious than the short and shallow tunnels at Audley End, Essex, imposed by Lord Braybrook's concern for his views.

Level crossings were frequent, however; and though the bridges built to accommodate roads over railways were often more impressive than those which carried the lines themselves the eastern railways had to construct some costly opening bridges as well as timber estuarine viaducts. That built by George Stephenson's colleague, G. P. Bidder, across the Wensum at Trowse in 1845 is believed to be the first railway example of the modern type which swings from a central island.

CHAPTER EIGHT

Uniting the Kingdom

The graph overleaf shows that the actual rate of railway building never exhibited any dramatic discontinuity, though there was a notable acceleration in construction after 1847. But the first column of the table above it illustrates just as clearly that something remarkable happened to railway *planning*—if so purposive a word can be used—around the middle of the decade. This 'something' was a sudden realisation of the profits to be made from railway speculation, and a corresponding over-valuation of such schemes by a lay public which for the moment was without other outlets for its investment cash.

By 1844, 2,236 miles of public railway were open in Britain (of which about 400 were in Wales and 500 in Scotland). But in the same year more than this whole length, itself a mere fraction of that *projected*, was authorised by a parliament which made only faint efforts to control the situation; and in 1846 more than twice as much new mileage was sanctioned in twelve months as had been built in a vigorous decade. In cash terms, about £70m had been spent on railways to date and work in hand accounted for some £7m more. But the 500-odd schemes which existed only on paper—and which in most cases could not have existed profitably, if at all, off it—represented an investment of some £500m.

Such were the conditions of the 'railway mania'. Its growth has been often and well described, though perhaps never better than by John Francis who wrote fifteen years later of the speculative frenzy which possessed—

> The noble who in the pride of blood and birth had ever held traffic in contempt . . . The priest who at his desk prayed to be delivered from the mammon of unrighteousness [but offered his] scrip at a premium . . . The lawyer, who, madly risking his money, sold the property of his client to meet his losses; the

Mileage table of Britain's railways 1823–51

Year	Mileage sanctioned	Mileage opened	Total sanctioned (Cumulative)	Total opened (Cumulative)
1832	39	26	419	166
1833	218	42	638	208
1834	131	90	769	298
1835	201	40	970	338
1836	955	65	1,925	403
1837	544	137	2,469	540
1838	49	202	2,518	742
1839	54	227	2,573	970
1840	—19	528	2,553	1,479
1841	14	277	2,568	1,775
1842	55	164	2,623	1,939
1843	90	105	2,717	2,044
1844	810	192	3,524	2,236
1845	2,816	294	6,340	2,530
1846	4,540	606	10,881	3,136
1847	1,295	740	12,176	3,876
1848	373	1,253	12,549	5,129
1849	16	811	12,565	5,939
1850	7	618	12,572	6,559
1851	126	243	12,698	6,803

The final column is presented graphically below

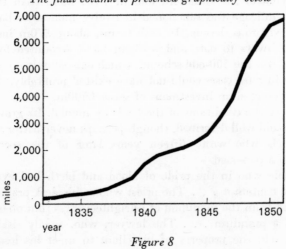

Figure 8

THE DEVELOPMENT
OF THE OVERALL ROOF
23 *Top* York
24 *Middle* Stoke-on-Trent
25 *Bottom* Newcastle

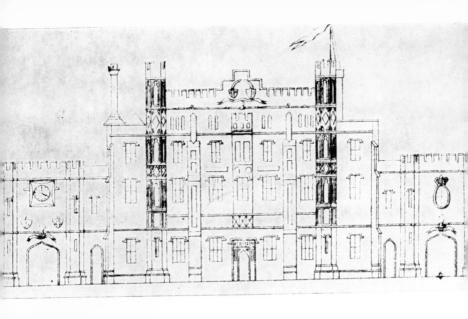

A WESTERN RAILHEAD   26, 27, 28 Bristol, Temple Meads

physician who perilled the savings of a life and the well-being of a family; the chemist who forsook his laboratory for a new form of the philosopher's stone; the banker who in the city and the senate denounced all speculation as illegitimate; the deacon of the meeting house; the warden of the church; the Jew, the Quaker, the saint, the sinner.

Some recession was inevitable and some fortunes were lost, so that soon afterwards George Hudson himself had only the tatters of worthless prospectuses and share certificates to keep away the cold. Overall, however, there was no violent market collapse but rather a healthy reversion to the more realistic conditions which asserted themselves in the 1850s. Certainly if the rate of railway building itself was affected by the mania and its aftermath these effects were delayed ones and (as has been noted earlier) were largely evened out by the cycle of conception/survey/parliamentary approval/commencement/completion. This cycle at best extended over several years, and in all cases was governed by the realities of the supply of experienced engineers and labour.

For two reasons, however, the watershed of the mania and recession is far from irrelevant here. In the first place, the building of several systems already described was interrupted by the failure of public confidence: the GWR did not consummate its designs on the Mersey until the mid-1850s, for instance, nor the Midland line attain Manchester until the early 1860s. (Similarly, many of the schemes described in the present chapter had been physically commenced before the upheaval.) The second reason is that 1848 marks the beginning of the end of the 'heroic' age of railway building.

There are economic reasons for this change of emphasis in the later 1840s: after the recession the financial prospects of a projected railway were examined rather more keenly, while at the same time the effect of the first railways themselves on the economy had been such that competitive fields of growth were now opening up. There are social reasons for it: the investment wave had itself done much to break up large fortunes and redistribute capital into smaller and more localised pockets. There are psychological reasons for it: the generation of Brunel and Robert Stephenson was certainly far from spent and had great achievements still before it, but to an increasing extent railway building was becoming a routine skill which could be entrusted to assistants working from precedent and textbook. There are topographical reasons for it: only a limited number of major land-dividers remained to be subdued, and these were in tough and underpopulated country where the highest standards could never be justified. Above all, and of the greatest relevance here, there are technical reasons for the change which came over the art of railway civil engineering before the middle of the century.

For this period coincides with a general slackening of ruling gradients on Britain's railways. For the first time in more than two hundred years there was a lesson to be learned from overseas—from North America, for instance, or from the Alpine nations of Europe which had now commenced their own railway systems. And this lesson was that gradients of many miles at 1 in 75 or steeper could be successfully worked even by the light locomotives of those times.

This is not to say that the rigid restrictions imposed by the Stephensons, Brunel, Locke, Rastrick, Cubitt and the rest on the nation's early trunk lines were mistaken: even with today's usage of electric and (to a lesser degree) diesel traction, the figures of 1 in 100, 1 in 200 and 1 in 300 represent real benchmarks of possible speeds and of the efficient use of fuel. But such criteria *were* those of a perfection which now had to be abandoned in the interests of economy; and though we have seen that the pioneers themselves had relaxed their standards over short runs where there was unusual topographical reason for so doing, the position after 1848 was that fairly protracted gradients steeper than 1 in 100 became regarded as acceptable in normal main-line usage. It was seen as better to spend a little more in running costs and much less in interest charges on an expensive building scheme—or, in the extreme, better to have a steep line than no line at all.

One other important change connected with the economic upheaval of the mid-'forties must be mentioned here, and this was the tendency for major companies (and particularly those which had a London railhead) to amalgamate openly into groups under new titles. It is true that well over half the hundred-odd railway enterprises at work in 1846 were mere creatures of larger ones, and true too that quite important, if localised, groups of a truly independent nature were still being formed. (Typical of these was Robert Stephenson's North Staffordshire railway, which was to grow up in 1848–9 around its headquarters town of Stoke-on-Trent.) But 1846 witnessed a number of those changes of state and style which, at least south of the Pennines, set the pattern for a regional division of Britain which was to endure for three-quarters of a century—and which can still be recognised, for all British Railways' attempts to impose a spiritless homogeneity, after a further fifty years.

Since this division forms the basic anatomy or five-finger exercises of the student of English railways, it may be presented here in the form of a count round a clockface whose centre is London and whose hour hand begins by pointing down the Thames at 3 o'clock. From there round to 5 o'clock was for the moment the undisputed province of the South-Eastern company. From 5 to 7 the London, Brighton and South Coast took over, and from 7 until 8 or after the London and South Western. The Great Western operated in a fairly narrow corridor around the 9 o'clock line, for at 10 the London and North Western—the successor of the Birmingham and Grand Junction companies—became dominant until 12 or beyond. The last quarter of the dial belonged to what was to be known as the Great Eastern railway.

This presentation, in itself reminiscent of the heptarchical England of the Dark Ages, is in several ways oversimplified: it neglects, for instance, the enormous 'head' supported by the GWR's spine, and the fact that at least two major companies remained to be slotted in. But it makes the point that there were some half-dozen frontiers along which the London-based systems alone could quarrel and interpenetrate. Away in the north the position was more complex yet, with companies such as the Stockton and Darlington still independent, and with the Lancashire and Yorkshire and the Manchester, Sheffield and Lincolnshire appearing as the rechristened proprietors of the Littleborough and Woodhead trans-Pennine routes respectively. Furthermore, three other major companies now worked out of Yorkshire—the future North-Eastern up to Newcastle, the future Great Northern down past Doncaster, and the Midland railway. This last had been formed by welding together the three lines radiating from Derby, a town which had been greatly extended (though not created) as the centre of railway engineering which it remains. With Hudson's force behind it the Midland had extensive interests: Gloucester and Lincoln, for instance, were both in its empire by 1845.

Here in the Midlands and north, too, the marcher wars and border raids were under way, while Scotland was shaping up for its own battles. But for a few years yet (and, in one direction, for a few *decades* yet) the railway engineers had real and important calls on their skills as they laid their lines on the face of a Britain beginning to show the signs of industrial middle age. In the south there were few important new developments before the middle of the century, though the SER commenced the fine direct line to Hastings which was opened in 1852 and among numerous other activities cannibalised the $2\frac{1}{4}$-mile Higham canal tunnel, with its rough-hewn portals and huge central shaft, in order to give access to the Medway towns. But away in the deep west Brunel was extending the GWR empire to the tip of Cornwall, then still a county of industrial significance.

Here there were two companies involved, of which the older was the West Cornwall railway which reached Penzance in 1852: this was originally of standard gauge but used a Barlow track which gave rise to frequent derailments. The broad-gauge Cornwall railway proper was directed towards Falmouth and took longer to complete. Both lines (and their branches) are notable for the way in which they pressed through difficult country on a very limited budget: the more western and inland one, for instance, had thirteen summits in its fifty-odd miles, frequent gradients of about 1 in 65, and correspondingly sharp curves. But their outstanding feature was the use which Brunel made on them of timber viaducts.

Wood was never a typical structural material for Britain's railways, though Jack Simmons describes some curious early examples of timber bridges, a few laminated structures (a fragment of one of which is preserved in the York museum) were built in the north-east, and the strutted *forms* of wooden bridges were now and then reproduced in iron. Certainly high trestles of the American type, or 'River Kwai' timber trusses, never appeared in Britain. Brunel's use of wood for both over- and under-bridges in various special situations at Bath, Bridgwater, Newport and elsewhere has already been mentioned (and can usually be accounted for by his dislike of cast iron) and he and others also built a number of timber viaducts crossing unnavigable creeks. But in the steep inland valleys of Cornwall stone would surely have been employed were it not for the need for economy.

As it was, however, Brunel was content to reduce capital costs by building for a life of from ten to thirty years: after this, he argued, more permanent structures could be built if the line prospered. He also adopted a prefabricated system and a routine of regular inspection and replacement, so that no life was ever lost by the failure of a timber viaduct.

Some thirty such structures were built between 1845 and 1860, all but a few South Walian examples being in south Devon or Cornwall. Typically, they were built on wooden or masonry piers up to an entablature some 35 feet below track level, from which sprung a fan of timber ribs supporting the deck. The overall effect, as often in Brunel's work, was curiously oriental.

These viaducts were up to 150ft high and 440 yards in length, perhaps the finest being that at Ivybridge. None survive, as would be expected of such deliberately temporary structures: in fact, it is surprising that the last saw service as late as 1934. But in some places in Cornwall—that county of lonely and mysterious relics—the piers of the original viaduct stand beside a later structure of stone and metal, as they do at Liskeard.

Brunel's greatest timber viaduct was planned to fit into place as the last link of his sunset chain of railways by crossing the Tamar at Saltash to unite Devon and Cornwall. By 1850 work had indeed begun on this crossing. But the Admiralty had insisted on a headroom of 100 feet to accommodate the tall-masted ships of the age, and it became clear that only a metal structure would serve. Financial stringencies meant that there must for a while remain a water-break outside Plymouth on the way from London towards the Lizard.

This chapter must now leap northwards, overlooking the earliest of those long cross-country branches which were typically linked to radial main lines at both ends and crossed others en route. For a century these added their charm and amenity to the English scene, but like the host of short spur-lines built to towns such as Bedford they rarely presented unusual engineering features. And this phrase can just be applied to another cut-off or 'relief' route backed by the LNWR companies and opened in 1847, the year after their amalgamation into that complex which was to regard itself—not without reason—as 'the premier line'.

Constructed under Stephensonian influence and after long political squabbles, this fifty-mile direct link between Rugby and Stafford avoided Birmingham and gave a more direct approach to Manchester. For the most part it followed the valley of the Trent and its lateral canal, and just before rejoining the older route it had to

er the handsome woodlands of the Earl of Lichfield at
ugh. This peer's ancestors had scattered some curious
ioilies about the estate; and these the railway company flattered by
adorning their tunnel with equally extravagant portals, castellated
on the west and Egyptian to the east. By an ironic twist, the great
house is now a county council museum of industry containing
railway exhibits. The whole area is worth more study than today's
100-mph transits of the Trent Valley railway afford.

In the north country itself, Harrogate (then England's most
favoured inland resort) was rail-linked in two directions, one of them
involving the building of the fine, mile-long Crimple viaduct and
Otley tunnels—the martyrs of which latter are commemorated by a
tomb in a nearby churchyard which reproduces one elaborate portal.
A seaside line which implied heavy engineering was that to Scarborough; and in the same area another cut-off railway, the Leeds
and Thirsk of 1849, was noteworthy for Thomas Grainger's Bram-
hope tunnel of some 2¼ miles. In building this some 1,500 million
gallons of water had to be pumped away.

A new Pennine crossing was also opened in 1849, that of the
Leeds, Dewsbury and Manchester railway. Like the Leeds and
Thirsk, this owed its commercial origin to Hudson's desire to build
up a self-contained empire—the future North-Eastern railway—
north of York; but there was sufficient traffic potential to justify
this more direct route between Manchester and central Yorkshire.
Its summit feature was the Standedge tunnel, originally (like
Woodhead) opened as a single bore but later augmented by first
another single track and then the double one which remains in
service. Although Standedge gave no particular troubles it is notable for two reasons—that at more than three miles it was slightly
longer than the 'old' Woodhead and so held the record for British
tunnels for more than a decade, and because it ran parallel (and was,
in fact, linked by cross-galleries) to a canal tunnel.

Despite such miscellaneous local works, the late 1840s were
dominated by the four great schemes, all begun before the recession
and among the first to recover after it, which give this chapter its
title. Of these, the formation of a 'west coast' route to Scotland by
prolonging the line through Warrington up to Glasgow had made
the greatest progress, for first Preston and then Lancaster had been
rail-served by schemes engineered by Joseph Locke. By 1846 a

through route already existed as far north as Carlisle, though with a
stiff climb of four miles at 1 in 75 over the 915 feet of Shap Fell on
the way. Beyond the border, too, the early lines in the Glasgow area
were being linked into networks which extended towards England by
way of such impressive works as John Miller's sandstone viaduct at
Ballochmyle, Ayrshire, whose semicircular central span 181 feet wide
and 163 feet high was for some decades after 1848 the greatest of its
kind in Europe.

But between rose a crossing of the uplands which was yet another of
the routes which George Stephenson (who to the south had favoured
a coastal trace crossing Morecambe Bay on a causeway so as to avoid
the heights of Lakeland) had dismissed as impossible, as indeed it
would have been had his canons been observed. Joseph Locke, though,
found the Annandale country not unamenable to his own 'up and
over' methods; and it is typical of these that the line had no major
tunnel or bridge and few engineering works of any type. It is
instead marked by the long 1 in 75 climb through Beattock which,
like Shap itself, laid a heavy load on steam locomotives.

So the first Scotch expresses—and trains, like whisky, mist,
shortbread and adhesive tape, are permitted to be neither 'Scots'
nor 'Scottish'—ran in 1848, soon after the echoes of the most
violent of the navvies' internecine punch-ups had died in the valleys
of Penrith. Meanwhile, on the opposite coast, the main line had
remained halted at Newcastle—though, rather surprisingly,
Scotland's own railways were here not only reaching south along the
coast to Berwick but had as early as 1846 attacked the tough country
inland.

Inevitably the continuation of the east coast route—that Great
North Road of railways which had challenged the visionaries for half
a century—was put in the hands of the Stephensons, and inevitably
too they counselled a level shoreline track. There had been one
extraordinary moment when Brunel had crossed the Tyne with a
scheme to build a more direct inland route which should be, if not
broad gauge, at least atmospheric. But (not surprisingly) he was
hooted out of the homeland of the conventional railway, and the
route decided on was one which avoided heavy earthworks at the
cost of constructing two great bridges.

In fact, even the reference to Newcastle above is a slight exaggeration; for though there were rail bridges higher up the Tyne the

most direct route to Newcastle from the south was by way of a ferry from Gateshead. Late in 1849, however, Robert Stephenson's historic 'high level' bridge was opened there.

The total length of this was (and is, for it is still in full service) 750ft; but it was broken into six equal spans, these being supported on sandstone piers. Restricted to cast and wrought iron for the upperworks where his colleague John Green had planned a laminated timber structure, Stephenson used these materials to build, not plain arches of the Coalbrookdale type which exerted a great lateral thrust on their abutments, but a series of 'bowstring' arches of the form which had appeared a few years before in this same countryside. This handsome type of bridge, in which the outward thrust is contained by tie-rods or the decking itself, was to prove popular until the age of mild steel; but the high level bridge at Newcastle is unusual in that it is double decked, with the level above the arch carrying three rail tracks and a road being slung below at tie-rod level. Such multipurpose structures were to become commoner overseas than in Britain.

At the same time work had gone ahead on the other great waterbarrier of this route, the crossing of the Tweed at Berwick on the Anglo-Scottish border. Here there were no navigational problems and hence Stephenson could adopt the most conventional solution and build, 126 feet above the tidal waters, a chain of twenty-eight arches of warm red stone. These were laid out on a gentle curve, and though somewhat in need of refacing still provide a splendid entry into North Britain. Their building involved a labour force of 2,750 men.

This Royal Border bridge, which followed a temporary timber structure, was opened by the queen in 1849 too. It was in its time described as 'the last act of the Union', though in fact the railways had entered Scotland two years before near Gretna Green and an inland route was soon to be added. A point of note is that its foundations were laid, on a firm base below fairly shallow water, inside cofferdams which had themselves been built round piles driven by John Nasmyth's newly invented steam hammer.

This hammer was shortly to be adapted to riveting by William Fairbairn, for steam was beginning to come to the aid of steam. But for by far the greater part a civil engineer in the middle of the nineteenth century still relied on the primitive 'tools' known to Brindley a hundred years before, such as the theodolite and gunpowder. Thus, a millionaire engineer personally supervising two such great, separated and simultaneous works as the Newcastle and Royal Border bridges would today demand the use of a private helicopter. But despite his continuing reliance on horseback for local transport, Robert Stephenson was in the same period of 1847–50 not only joining England and Scotland but also working on an England–Wales–Ireland scheme several hundred miles to the west. It is not surprising that a man in his mid-forties, faced with such a weight of work and responsibility, had long before become a heavy drinker, a chain smoker and probably a laudanum addict. What *is* surprising is that he lived to complete another decade of such labours, mainly overseas.

In 1847 the North-Western system still reached little further west than Chester—an unhappy city for Robert Stephenson, since in that year the skewed bridge across the Dee where he had recently used cast iron in spans nearly 100 feet long collapsed with the loss of five lives. (These pioneering giants who worked by instinct as much as by mathematics *had* their failures, and others were not as success-ful as Brunel in hushing them up.) But the elder Stephenson had surveyed a continuation of this line to Holyhead, which since Tel-ford had built his highway had become the established port for Dublin and most traffic to Ireland; and with the revival of railway activity his son again took over. The route was typically Stephen-sonian in that it was coastal (involving the construction of not only sea walls but avalanche sheds in the area about the Penmaen head-land) and preferred river-crossings to the surmounting of crests. But in fact the only major estuary was that at Conway, and the leap across the Menai straits to the island of Anglesey was unavoidable unless, as was suggested, a new port was to be created.

Had Conway been the only bridge on the line Robert Stephenson might well have used a bowstring structure there, though the city fathers insisted that anything erected must look medieval. But in fact he was able to treat the crossing of this unnavigable estuary almost as a test-bench for whatever was to be used to span Menai itself. And there Stephenson was faced with a problem which rail-way engineering could no longer avoid—the fabrication of bridge spans whose length was measured in hundreds of feet. For Admiralty orders similar to those which were troubling Brunel at Saltash in-sisted that no piers could be erected at Menai other than one on the small but central island which named the bridge, the Britannia rock.

The obvious solution would have been to build a suspension bridge similar to those with which Telford had carried his road across both Conway and Menai itself several decades before. But since a suspension bridge built to extend the Stockton and Darlington line over the Tees to the railway-created port of Middlesbrough had failed within a few years of its opening, British engineers had decided that the suspension principle was unsuited to bridges which had to bear the heavy live loads of rail traffic.

The story of Robert Stephenson's solution of the problem is one of the often-told classics of engineering history. Beginning in 1845, he and Fairbairn launched a series of experiments on the strengths of wrought iron girders and of the rolled plates which had been developed after 1830 for the building of locomotive boilers: a mathematician also advised. After nights when 'he would lie tossing and seeking sleep in vain . . . the tubes filled his head . . . he went to bed and got up with them', Stephenson conceived the plan of running the twin tracks through rectangular tubes, and built a sixth-scale model. After extensive trials and corrections with this he went ahead with the structural work for Conway.

The girder here was 425 feet long, though so low above the water that it has today been shored up with timber struts; and so it provided a full-scale prototype on which Stephenson could measure deflections. Fully satisfied, he continued with preparations for Menai itself: this had two approach spans of 230ft and two main ones of 460ft supported more than 90ft above the water, whereas the longest such girder previously fabricated had been little over 30 feet long. Another striking feature was that the rivet-holes were drilled by machines controlled by punched cards. Up to the last the consultant mathematician had advised that the great tubes should have some of their 1,500-ton weight taken by cables—on the principle, as it has been said, of wearing braces as well as a belt—and although Stephenson won his battle to leave them self-supported, pavilions were built at the tops of the towers to contain the projected cables.

Other features of a bridge whose general appearance was due to Francis Thompson were the portals inscribed ROBERT STEPHENSON and their thirty-ton guardian lions; these were designed by the railway sculptor John Thomas who was also responsible for such compositions as the pediment at Paddington, and were pleasantly described by a contemporary as 'of the antique, pimple-faced, knocker-nosed Egyptian kind'. A plan to crown the work with a statue of Britannia was wisely abandoned, however.

So, in Robert Stephenson's heroic year of 1849, a fourth great bridge was opened to traffic. But bridge building is not merely the most glamorous form of civil engineering. It symbolises a universal human desire for union, and it is not for nothing that the pope is anciently *pontifex maximus* or that three thousand years earlier the Egyptians believed: 'The mitre of mystery and the canopy his/Who darkened the chasms and domed the abyss'. In the middle of the last century Robert Stephenson took perhaps the greatest single step forward in this art since the Coalbrookdale use of iron by devising those tubes which were among the most sophisticated forms of construction possible before the coming of mild steel. Yet it was less than forty years since a Scottish tramway had first crossed a river on an iron deck clumsily cantilevered out from the side of an older stone road bridge.

For more than a hundred and twenty years more this object of service and beauty dignified North Wales, though British Railways showed a typical disrespect for the past by letting lapse the tradition of keeping bright the last of its two million rivets. (This rivet, though, had a less fiery career than those which were tossed white-hot through the night from the shores of Wales to the girders high above). Then, in the summer of 1970, a fire started by two young vandals trespassing on the line spread to some timberwork which curiously lined the tubes. Although the latter appeared undamaged, British Railways claimed that they were weakened or distorted beyond repair. It may not be irrelevant to note that the authority had in any case recommended the singling of the line.

As this book went to press the future of the greatest of all relics of early-Victorian civil engineering was in grave doubt: there appears a real danger that the historic tubes will be surreptitiously dismantled and replaced by some gimcrack and asymmetrical structure epitomising our age of public squalor. In fairness to British Railways, that authority cannot be expected to add to its other responsibilities those of a preservation trust. But there would be no difficulty in raising the funds needed for a proper rehabilitation of the Britannia bridge, and BR must not be allowed to add this to their long list of recently-destroyed national monuments.

In 1850 Stephenson opened another tubular bridge, with a 237ft span this time, at Bretherton in Yorkshire. This work, never renowned, has now vanished. But it serves to return the story to the north, where one more link in the mesh of Britain's railways was falling into place.

The construction of a route from London to York was (as we have seen) one of the earliest of railway dreams; but in fact even Edinburgh had been reached with York still having only a devious approach from the south. Hudson had an interest in the old route through Derby; and if anything more direct was needed, he had vehemently argued, it could be supplied by a link from Doncaster to the East Anglia of which he was also master. After much debate as to whether a direct crossing of the Fens via Lincoln or a 'towns' line running by way of Grantham and Retford should be adopted, parliament had given the latter scheme priority and in the mid-'forties a railway was hence planned to approach Peterborough from the north.

From Peterborough to London was less than ninety miles, and a **direct** line here would provide by far the best access from the capital to York and all points north. The scheme was, of course, even more violently opposed by Hudson. But in 1846 he lost his battle, and a new and independent company, the Great Northern railway, was authorised to build the London to Peterborough link and operate up to Doncaster. Its engineer, appointed in dramatic circumstances, was the builder of the Dover line—William Cubitt, who was assisted by his son Joseph and by another member of the family, Lewis. Its main contractor was Thomas Brassey.

Work was interrupted by the economic upheaval, there was scant traffic to capture north of the London suburbs, and though the country was so level that there was little difficulty in keeping the ruling gradient down to 1 in 200 (a stringent figure for the late 1840s) and the curves correspondingly relaxed, this last and longest of lines built in the manner of the pioneers presented features of its own. Thus the crossing of the Nene just south of Peterborough is interesting in that one of the original iron spans still carries traffic. Further south again lies Whittlesey Mere, a three-mile-wide 'quaking bog' which (Brassey wrote) 'you could stand upon and shake an acre of it together': this was conquered by the Chat Moss technique of sinking brushwood rafts. And at the far end of the line the twenty-five miles into London implied the building of nine tunnels (some, like Potters Bar, fairly long, but all shallow and through clay) and the fine Digswell viaduct. Nearly 100 feet high and 1,520 yards long, this last still rises above Welwyn on its forty brick arches.

The Great Northern was essentially completed by 1850. though two more years were to pass before the opening of Lewis Cubitt's splendid station at King's Cross, which with its twin arches marking the arrival and departure sheds behind them remains the purest work of 'engineer's architecture' in London, Britain or perhaps the world—and whose high platform levels probably witness the final steep plunge of the line to burrow below the Regent's canal. But 1850 (or better, perhaps, the following year of 1851 when at the Great Exhibition from which King's Cross acquired its clock Britain showed the world the fruits of an industrialisation in which the railways had played the leading part) marks more than the fulcrum of that fantastic century which is only now beginning to be seen in perspective.

In the previous five years railways had come of age, a fact which shows alike in the advances in locomotive design dealt with by John Snell and in such details as the evolution of the central lever-box controlling points and signals—with the latter perhaps being grouped on iron gantries spanning the tracks and in any case now separated from the box itself. In all, some £250m had been spent on a network of nearly 7,000 miles of railway; and the larger companies, in particular, were marked by that quality of professionalism which ever since Liverpool and Manchester days had impressed a public which contrasted their para-military smartness with the seediness which had characterised many aspects of the stage-coach era.

But at the same time, as was noted at the start of this chapter, the heroic quality was passing from British railway building. Of its nature the pioneering age could not last for more than a few decades, and with a handful of important exceptions the great challenges were now to be found overseas. And so the year of 1851 may be as good a point as any to break the chronological flow of this book for one chapter, and to look at what had been achieved through the eyes of the architect rather than of the engineer.

CHAPTER NINE

The Search for a Style

There is no clear division between the arts of civil engineering and of architecture, and many of the railway structures mentioned previously in this book—the portals of Shugborough, the pavilions of Balcombe, even the towers of Menai—display more conscious adornment than do comparable works of the canal age. Aesthetics even entered into the treatment of iron bridges in sensitive situations, such as the late one at Derby Friargate with its delicate iron work by Andrew Handyside or those on the Dulwich estate designed by Charles Barry. But the railways also demanded and begat a host of structures which *as a whole* were closer to the traditional province of the architect than of the engineer.

At one end of the scale were the buildings which had direct ancestors in canal and tramway technology—the small and perhaps residential office block which derived from the weigh-house, the lengthman's isolated cottage, or those plain and massive warehouses which even towards the end of the nineteenth century recalled Telford's proportions. At the other extreme, the influence of the railways became so widespread (and one sometimes feels that by the middle of the century they were beginning to run the country rather than it them) that it is hard to say where railway architecture ends.

Company towns in the strict sense, such as the Swindon and Crewe which themselves had a precedent in the canal age Stourport, have already been touched on; and these were followed by later 'model works' (such as Horwich, Lancashire) of interest to historians of industrial architecture. But there were also railway towns in another sense, those which grew up round a traffic centre and were largely influenced by the architectural mannerisms of its main company even when they did not stand on its land and depend on it for their entire prosperity. An early and 'pure' example of this type of town is the Stockton and Darlington's creation of Middlesbrough: a more complex case is the way in which the centre of Stoke-on-Trent was transformed by the North Staffordshire company. Nor was this influence felt in industrial centres only; for just as Brighton had been created and Bath transformed by the stage-coach era, so the railways built up their own watering-places such as Saltburn, Cleethorpes and Bournemouth in the 1850s and 1860s and were responsible for the growth of many other seaside resorts and inland spas. Finally, early in the present century, a railway company became directly interested in the purely residential development of 'Metroland' in the Chilterns.

When to this complex of ideas is added the effect of the railways in cheapening the transport of Welsh slates and Midland bricks, and so in homogenising the architectural scene of the nation, it is clear that even the briefest survey of 'The Interrelation of Railways and British Architecture, 1830–1900' would be outside the scope of this book. The subject has, in fact, recently received full-length treatment, and this chapter is intended mainly to summarise the early development of that archetypal structure, the passenger railway station.

Here there was no historic line to develop from; for the canals had never greatly concerned themselves with handling human cargo and the stage coaches could make shift with inn yards for their marshalling of a maximum of a few dozen passengers at one time. The station was not, though, the only completely new type of building demanded by the railway age; for even if the traditional signal-box (itself now so swiftly vanishing that one has been rebuilt inside the Science Museum) can be argued to have an ancestor in the toll-houses and lookout posts of the waterways, the engine shed owed nothing to the canal stable block.

In its polygonal or circular 'round-house' form this was, in fact, the purest innovation of all railway architecture, and demonstrates how early and confidently an unusual technical problem was solved. The most famous and interesting such building, and probably the earliest too, is the engine shed which Robert Stephenson built in 1837 at Camden Town at the top of the incline into Euston; since this method of working was soon abandoned the round-house itself was disaffected from railway use before its original structure had been tampered with, and after serving for a century as a gin warehouse is now used as a theatre. In the next year an octagonal structure was built by the Croydon railway at New Cross, but even this form proved unjustifiably costly. Several other round-houses were built in provincial centres, however, and a few may still remain in railway use.

Human beings could not be marshalled as neatly as machines; and the evolution of a satisfactory plan for a passenger terminus was, as we shall see, to take a considerable time. The ordinary wayside station, on the other hand, came to birth apparently painlessly. But before discussing it there are two general points to be considered.

The first concerns the frequency of siting of such minor stations. Early lines varied in their ideas on this subject if only because the traffic potential of the country through which they passed itself differed; but there was perhaps a tendency for the first railways to be conceived more in the 'inter-city' terms of today than with the hopeful mid-Victorian idea of 'opening up the country'. (A similar progress had taken place in canal routing, with the earlier arteries running from A to B but many later canals being promoted with vague prospects of stimulating local development.) Thus the six-mile Canterbury and Whitstable railway had no intermediate stations, and on the London and Birmingham their average interval was nearly seven miles. In high-Victorian times, by contrast, the national mean became a station every three miles or so.

On even the least changed of Britain's railways one would, for this reason alone, rarely find a series of stations built contemporaneously: rather, the older and newer would be found interspersed. But the actual situation is usually more complex still. Track layouts have been altered, towns have swollen and shrunk (the general rule being that the trunk routes of the 1830s and 1840s crossed the most prosperous country and hence had stations which were early outgrown), and even before the present century some of the older stations had become irrecoverably decrepit. Hence the age of a railway itself is little guide to that of the buildings—as opposed to engineering structures—which will be found along it, and four successive 'stations' on a typical British main line may consist of a little-altered 130-year-old stone building with later additions on the facing platform, a brick rebuild dating from the 1880s and about to make way in turn for an aluminium prefab, a concrete box of the 1930s, and an abandoned site.

MAJOR PROVINCIAL STATIONS

29 *Top* Chester
30, 31 *Middle and Bottom* Shrewsbury

STATIONS—THE CLASSIC CONCEPT

32 *Top* Curzon Street, Birmingham
33 *Middle* Monkwearmouth, Co. Durham
34 *Bottom* St Marks, Lincoln

This leads to a second introductory point. This book has attempted throughout to make clear where a structure has vanished; but Britain's stations are now being closed or rebuilt at a rate of a dozen or more every year. Furthermore, the victims are sometimes demolished, sometimes left to the depredations of weather and vandals, and sometimes converted to new purposes—these ranging from the expected farm cottage or repository to a boathouse, a sculptor's studio, a youth club and even a chapel. Hence any information as to present states must be treated with reserve.

Time has dealt particularly harshly with the intermediate stations on the pioneering trunk routes, for nothing remains of the originals on the Birmingham railway except part of Watford. Mocatta's smaller stations on the Brighton line, too, have vanished save for fragments at Three Bridges and Horley, and of Tite's for the Southampton railway only Micheldever and Eastleigh survive. Most regrettably of all, the last two examples of Brunel's village stations east of Bristol were deliberately destroyed only a few years back. A number of slightly later Brunel stations, however, survive deeper in the west as at Exeter (St Thomas) and Torre. Those associated with the atmospheric scheme have already been mentioned, but almost equally characteristic are those where a gable-ended 'tent' of wood and glass covers the whole track and platform area.

This was a local variation, though, as was the habit found occasionally in Kent, East Anglia and elsewhere of staggering or off-setting the up and down platforms. For soon after 1840 a standard layout of the small station could be found in every quarter of Britain. There was a substantial block, probably on the 'town' side of the tracks, with booking-hall, offices, toilets and more or less generous waiting accommodation segregated by class and sex: the station-master lived above, or in more important cases had a separate house in the same style (which has sometimes survived the station itself) with his assistant using the flat: and there would be various matching out-buildings. These were sometimes built to a higher standard than their site and purpose would seem to justify, as when a rural carriage-shed was furnished with finely carved corbels. The platform—lower than today's, but even when a detached boardwalk more clearly defined than in overseas practice—was usually sheltered by a canopy which might show highly individual craftsmanship not only in the woodwork of its valencing but in the cast iron supports which had to be set far enough back to avoid swinging carriage doors. On the opposite platform amenities were at first usually minimal, though in the slightly larger station this might be treated as an 'island'. Subways and footbridges were initially rare.

This general plan, which still serves the needs of the majority of small stations overseas as well as in Britain, had no precedent in the taverns and wharf-houses of the tramway era. It was unknown on the Stockton and Darlington, and apparently on the Liverpool and Manchester, too. But it can be seen fully formed in the buildings of what has already been noted as a remarkable example of a line pre-served in what is close to its original state, the Newcastle and Carlisle railway. For at least half a dozen stations on this (including, suitably, Wylam) date from the late 1830s.

It would hence be of the greatest interest to know their archi-tect, for almost certainly one man was responsible not only for the basic plan but for modifying it to meet the needs of a dozen special situations. (He may also have been responsible for the Carlisle terminus in London Road. This is still in use as a goods depot, and indeed the freight station of many a middle-sized town is of interest as being its original passenger terminal.) But so far his identification has defeated dedicated research, and all that can be said is that he was *not* Robert Stephenson (though the present author has unhappily helped to spread that legend), is most unlikely to have been Francis Giles, but *might* have been either John Dobson or the important but little-known Newcastle architect, Benjamin Green.

The authorship of the Newcastle and Carlisle stations is one of the two great mysteries of railway architecture, the other being the identity of the designer of the astonishing 'baroque orangery' of 1848 at Newmarket, long disused as a passenger station but still pointing the moral that the poorest companies often built the grandest stations. But the Newmarket architect was an isolated folly-designer, whereas the designer of the Newcastle and Carlisle stations must be accounted of enormous influence.

Such wayside stations, often falling almost by accident on the out-skirts of communities where hitherto little of even local moment had ever occurred but which thereafter were to have an advantage over their neighbours, are the parish churches of the railways. Like churches they may sometimes have had no more of an 'architect' than a local mason working within an agreed formula: like churches they are frequently sensitive indicators of cultural and geological frontiers: like churches they have often been organically modified but occasionally present the all-of-a-piece jewel: and Sir John Betjeman has even traced a series of parallels between the elements of church and station. Perhaps the greatest resemblance, though, is that the 'parish station' proved amenable to the incorporation into a basic plan of a host of local materials and details, and that these were handled as felicitously as Brunel and Stephenson handled their grand tunnel portals.

The Newcastle and Carlisle stations were mainly of stone and adopted that small-manor-house Tudor style which three hundred years before had proved capable of fitting into very varied landscapes. Perhaps because the age of the Georgian brick box which had provided a formula for Telford to work from had ended just before the railways appeared, Tudor was to remain the prevailing idiom for at least the smaller station throughout the 1840s and 1850s. Many admirable buildings in this style remain, often designed by local architects such as William Tress (Battle, Sussex, 1852), Frederick Barnes (Stowmarket, Suffolk, and other gabled stations in that county) and perhaps Sancton Wood (Great Chesterford, Cambridge-shire, 1846). The half-timbered vernacular which a century later become debased into 'by-pass Tudor' was a rarer variant, but was used on the Oxford–Bedford branch, complete with barge-boards at such stations at Woburn Sands. Another localised style (unkindly called 'Victoriabethan') was that of the North Staffordshire company.

Everywhere, though, the most sumptuous such buildings were those likely to be patronised by royalty (Windsor Riverside, designed in 1850 by the prolific William Tite and originating from a smart deal by Prince Albert) or by the more abundantly-landed aristo-cracy (the demolished Trentham, Staffordshire, and Brocklesby, Lincolnshire). Stations of this type were often permitted only on condition that certain trains should or should not stop at them. To counterbalance such luxuries the unmanned, wooden-shack 'halt' or 'platform' also made an early appearance, and was followed by the big but bleak station designed to serve a racecourse or other excur-sion centre.

The largest and perhaps finest of Tudor stations is Tite's Carlisle Citadel (1847). This style was not unchallenged, however; for even before Brunel experimented with his oriental modes John Foster (who was probably the first railway architect of all) had built a Moorish arch over the Liverpool and Manchester railway. Such exotic vernaculars generally proved short-lived; but there was one important exception.

This was the style which became immensely popular around 1840 for not only railway stations but small houses in outer London and in residential towns and spas. It became known as 'Italianate'; but some of its most characteristic *motifs*, such as the pierced barge-boards and drip-mouldings, would have caused Palladio considerable surprise and its architectural history is complex. Its popularisation is often credited to Charles Barry; but there is some evidence that the plans of David Mocatta and others were themselves adapted for residential villas, and Italianate certainly became known as 'the railway', or even 'the commonsense', style. These stations are today harder to find than Tudor ones, since they generally served built-up (and hence rebuilt) areas; but a few fairly pure examples remain in, for instance, north London. The style penetrated to such remote parts as Gobowen, Shropshire, and can even be found reimported to the Continent.

Two other groups of small stations worth special note are the late-built but very vernacular granite ones of the Scottish highlands and the few which borrowed an existing feature—a genuine Elizabethan manor house at Bourne, Lincolnshire, a hollow tree in Herefordshire, or occasionally and temporarily an inn. Early Gothic is never represented in the small station, or even the larger one at this date: Norman appears only in details: and though classical influences became more widespread after 1845 pure greco-roman idioms were at first adopted only by one outstanding railway architect.

This was Francis Thompson, who showed a versatility surprising even among his peers since his stations frequently varied in basic concept rather than mere detail. Thompson built two series of stations for the Stephensons, one for the North Midland railway in the early 1840s and the other for the Chester and Holyhead line at the end of that decade and after he had published a work titled *Railway Architecture*. Of these later stations the main block of Holywell, Flintshire, is most notable, though other examples survive: from the former, there remain the disused Ambergate (where a section of the later junction station is also of interest) and, again disused, the uniquely greco-roman Wingfield, Derbyshire. It is typical of our false contemporary values that anyone who destroyed a canvas by the comparatively minor artist 'Wright of Derby' would rightly be accounted a vandal, whereas these gems by his greater townsman are being left to crumble away without protest.

While in this Peak countryside it is worth trespassing outside the present period to follow the North Midland line on its way to Manchester. Here two other Derbyshire stations are of particular interest—the original Rowsley, which may have been designed by Joseph Paxton before the Great Exhibition year, and Buxton, where the same genius of a gardener turned engineer-architect probably advised some fifteen years later. The latter is of additional note in that it marks one of the few cases where a local authority imposed a joint composition on two rival companies: the present age has typically proved less sensitive, and has permitted only a fragment to survive.

Rather curiously, though Mocatta had experimented with pre-fabricated components and fixed modules for his station buildings more than a decade before and Brunel himself had added the towers to the Crystal Palace, that epoch-making structure had almost no influence on the building of the *smaller* station until a round century had gone by. An exception is the former LNWR terminus at Oxford, now used as a goods depot, where the roof was bolted up from Paxton-type elements.

Turning from the parish church to the cathedral—or, perhaps even more appositely, from the gate lodge or gamekeeper's cottage to the big house—one must again consider both plan and adornment, function and style. And planning itself must be looked at under two heads, that of the major 'through' station (probably a junction where *some* trains originated or terminated) and that of the terminus proper.

Perhaps the most notable point concerning early stations of the former type was that several companies considered it easier to switch their trains about than their passengers, and hence built 'single-sided' stations. Inevitably these led to operating difficulties, and hence the only example of the arrangement which survives is at Sancton Wood's arcaded Cambridge (1846). But it was also adopted —with the modification of physically separating the two 'sides'—at several places on the Great Western system.

A fine example of a more conventional large through station whose exterior has changed little since its early years is Francis Thompson's Chester of 1848, whose thousand-foot-long frontage of classical form relieved with Tudorish vertical accents recalls Barry's compromise at the Houses of Parliament. For these more substantial stations, though, the purer classical orders came into their own, so that in several northern cities the finest building is a station boasting a noble Greek-revival portico or colonnade. Outstanding examples can be found at the Corinthian Huddersfield (J. P. Pritchett, 1847) and Hull Paragon (G. T. Andrews, 1848), whilst another member of the group is Dobson's Monkwearmouth (Co: Durham). A far more impressive building than the town seems to warrant, this last was erected *ad majoram gloriam* George Hudson and commemorates the 'railway king's' election to parliament in 1845. Ironically, it is now semi-ruinous.

From the public's viewpoint, perhaps, the distinguishing mark of this type of station was that it provided refreshment facilities such as were first introduced at Birmingham in 1837. This subject, though a little remote from that of railway civil engineering, has already been mentioned in connection with Swindon and Wolverton; but it may be that there is something still to be studied in the inter-relation between railways and British catering in the 1840s. It has recently been suggested, for instance, that the bar for serving food and drink (and with it the barmaid) was, if not invented, at least introduced to a normally-leisured Britain at such stations.

Swindon was also celebrated for its hotel, which was rather oddly designed with bedrooms on both sides of the tracks and linked by a bridge. More conventional hotels were built at other important junctions and terminals, either integral with the station offices in 'fore-buildings' making up an architectural unit (as at G. T. Andrews's Italianate station of 1851 within the walls at York and several London terminals) or close by and matching in style: the latter plan was followed at the first of all railway hotels (1837) at Euston, and later at King's Cross. A detailed study of such buildings would again be outside the scope of this book, but it is worth mentioning that the world's oldest active railway hotel is Thompson's detached structure at Derby (1841), where his 'trijunct' station itself was soon substantially rebuilt.

So at last one approaches a consideration of the British terminal station itself. The adjective of nationality is worth stressing, since the larger British station frequently differed from its continental and American counterparts in being set on the edge of its city. Early prints, indeed, illustrate stations such as Oxford standing without the walls in truly medieval fashion, surrounded by pigs and peasants; and this geographical remoteness, though due mainly to the relatively high land values which implied that the last mile into a city centre might cost as much as half a million solidly Victorian pounds, at least had the advantage that remarkably little interesting urban building was ever erased to make room for the railways.

Since the peripheral station is normally a through station, the pure terminus consisting entirely of dead-end bays (such as Norwich) is rare in Britain in even larger centres: compare, for instance, the railway plans of Manchester and Vienna. Even stations which began life in this form were often rearranged later to provide new links or to avoid the need to reverse trains, and in British practice the word 'terminus' is hence commonly used functionally rather than structurally and is applied to any station (like Edinburgh, Waverley) at which a majority of main-line trains originate, terminate or are remarshalled.

The world's oldest passenger terminal is also the world's oldest railway 'station' of any kind, if that term can be used of the cottage by a level crossing in St John's Road, Stockton-on-Tees, which opened its books in 1825. (It did not sell tickets: that innovation belongs to the Newcastle and Carlisle railway, and the *guichet* did not appear until 1847.) But even the briefest glance reveals that this building had hardly advanced from the canal wharf-house, whereas by 1833 the same railway had built a station at North Road, Darlington, so suited to its purpose that it still handles passenger traffic. Hence, between 1825 and 1833, a workable form had been devised for at least the smaller terminus.

From the two historic lines opened in 1830, the old railhead at Canterbury has long since vanished. So have the successive Liverpool stations whose arrangements were in any case unusual. John Foster's terminus for the L&M in Liverpool Road, Manchester, on the other hand, is still largely intact—demoted to a goods depot like many of its compeers, but like them viewable on application by the serious enthusiast. And it is this which represents the first terminus of a 'modern' type, if not the oldest of all true stations.

The trunk lines, however, had to handle traffic on a far larger scale; and it is clear that for a few years after 1840 the railways were still baffled by the problems of marshalling their travellers, trains and luggage. The usual layout was to have a 'train-yard' with a single departure platform on the left side (as approached by road) and a single arrival one on the right. Passenger amenities were grouped along the departure platform, the arrival side generally lacking even a barrier since many railways adopted the curious practice of stopping their trains just outside the terminus to collect tickets. Between these two platforms there might be as many as a dozen 'stabling' tracks.

Despite an analysis of terminal practice published by a logical Frenchman in 1846, it was at least a decade before the advantage of a grid of 'finger' platforms—or even of easy connection between the two sides by way of a cross-platform at their head—was appreciated, and well into the 1850s 'Platform 1' was considered the best site for a booking office. But the layout proved as chaotically unsuited to its traffic as that of the average provincial bus station of today. After an initial period of exaggerated respect for locomotives, passengers began to swarm across the tracks as across an inn yard. And inevitably the companies in their turn overcompensated, attempting to keep travellers under the type of lock-and-key surveillance today encountered at Spanish railway stations—and at airports everywhere.

It is not surprising that coherent architecture rarely emerged from such a situation, or that in large cities the termini of the early 1840s were often built as temporary sheds. But some kind of grand gesture had to be made if only to impress the stockholders; for it has been pointed out that, whereas the companies tried to project a friendly neighbourhood feeling through their country stations, their aim in city centres was to impress with a sense of hierarchy and solidity. The solution, which attracted the scorn of architectural critics but which in a time of experiment had a logic of its own, was to invest the *approach* to the terminus with a monumental grandeur.

Thus Paddington boasted an open colonnade long before Brunel commissioned Matthew Digby Wyatt to add an architectural gloss to his own basic plan and produce what is essentially the present station. Even the London Bridge complex, which never achieved maturity, began with a campanile. But the outstanding example was Euston, where the series of linked pavilions once constituted one of the finest if most useless sights in Europe. Today even Philip Hardwick's central Doric prophylaeum has been wantonly destroyed, together with the station buildings which were constructed by his son on as grand a scale a few years later. All there is left to preserve is the forebuilding of the corresponding terminus (now demoted to goods depot status) at Curzon Street, Birmingham, whose fine Ionic portico of 1838 has somehow survived amid very drab surroundings.

Of such early provincial railheads there are several other dignified semi-survivors. The 1840 terminus of the Preston railway in Penny Street, Lancaster, has become a nurses' home, whilst William Tite's Gosport (1844), like his turreted Bricklayers' Arms in south-east London, is ruinous and only the façade remains of his Southampton Terminus. Barely a fragment, too, subsists of Glasgow, Queen Street; and the fate of Cubitt's Dover Town is uncertain. Brighton still stands on its bluff above the town, though Mocatta's original work is now hard to see. But the Greenwich railway's terminal of 1840, though shifted *in toto* by a few hundred yards and converted to a through-station, has been little altered; and above all there is the case of Brunel's Bristol, Temple Meads, of 1841.

The whole threatened complex here is interesting since the forecourt, though dominated by Wyatt's felicitous extensions to the original 'Tudorbethan' work added after 1865, also preserves Brunel's pretty if smaller Perpendicular buildings. Its outstanding feature is the old train-shed with its 72ft wooden roof—wider in span than Westminster Hall and (according to Jack Simmons) the finest example of all early railway architecture now that Euston is lost. In the last few years even this has been disaffected from railway use: typically, it is now used as a car park.

By the climacteric date of 1845 large stations were being built with some confidence. But one component was still missing, a satisfactory roofing system. The uncertain climate of Britain favoured an overall glazed 'umbrella', admitting plenty of light, such as had first appeared on some fairly small York and North Midland stations designed by G. T. Andrews. But such roofs were limited in span so long as they had to be built up from short, straight members of iron or even wood.

Laminated timber was used in roofing Lewis Cubitt's splendid King's Cross as late as 1852—and still later at Copenhagen—and the iron shed in its various forms became as common as it often appeared mean (though there are good examples such as Stoke-on-Trent). But a notable example of an engineering advance producing a whole new type of architecture dates from the invention in 1849 by John Dobson of Newcastle—perhaps the greatest of all railway architects—of a method of rolling the long, *curved* iron sections from which true arched roofs with spans exceeding 100ft could be easily and safely built up. For a few mid-century years (and for a certain type of building which included exhibition halls, commodity exchanges and the like as well as stations), architecture was linked to technology as it had not been since the glories of the Gothic—though William Morris could never see the parallel.

The first great station to take advantage of this technique was an outstanding one by any standards. This was Newcastle Central, which if built as originally designed by Dobson would have become one of Britain's noblest buildings. Reality imposed economies and the appointment of the less ambitious Thomas Prosser; but the exterior is still very fine, with its classic portal completed in 1855 being a worthy commemoration of the importance of Newcastle on the railway scene. Dobson's own three-aisled train hall of 1850 is notable for being built on a sharp curve—an amazing extra difficulty, imposed by a site which also led to Newcastle's originally being one of the last of the single-platform stations, to challenge at the start of a new technique. Later the Crystal Palace building of 1851 popularised not only the great metal-and-glass roof as such but its use in a transepted form such as appeared at Paddington in 1854: here the transverse aisles were built to house the traversing gear which provided an alternative to the plethora of turntables, inherited from colliery practice, which was used in an age when trains were commonly marshalled in the terminus itself.

This account has now carried railway architecture down to that date, just after the mid-point of the nineteenth century, which has been reached by this book as a whole. It is not true that worthwhile station design ended shortly afterwards with the coming of the academic 'battles of the styles', for such fine and individual major stations as Prosser's new York (1877) were still to be built—as was St Pancras in London. A good smaller station was the now-abandoned Bath Green Park of 1870; and whole new schools of building were yet to appear, as exemplified by the rather debased stock-brick Italianate of the London suburbs and the French-renaissance, wedding-cake chateaux of the new south-of-the-river company terminals. Of these,

E. M. Barry's attractive Charing Cross (1867) survives, though not his even better Cannon Street.

But, once again, what had begun as an experiment or adventure had within a decade or two become a mature art; and certainly it is the stations large and small built before (say) 1855 which are of the greatest interest. The following list, which is largely based on David Lloyd's work, suggests some three dozen more of Britain's stations particularly worth study: Acklington; Ashby-de-la-Zouch; Atherstone; Audley End; Bath (Spa); Burton-on-Trent; Bury St Edmunds; Castle Howard; Cheltenham (Lansdown); Cromford; Dawlish; Earlestown; Ely; Fenny Stratford; Foxfield; Frome; Gainsborough (Central); Hampton-in-Arden; Hereford; Kettering; Littleport; London (Fenchurch Street); Lincoln (St Mark's); Long Marston; Louth; Matlock; Needham Market; Sandon; Scarborough; Shrewsbury; Spalding; Stamford (two stations); Swaffham; Stone; Thurston; Tynemouth; Wateringbury; Wemyss Bay.

All the above are believed to be still standing, though not all are used for passenger traffic. But many are threatened, as many of equal merit have already been lost—ironically so in an age when industrial architecture has become a fashionable subject. For despite the work of the Victorian Society the preservation of early stations is typical of all such movements in these overcrowded isles. Authority in its secret ways encourages the drift towards philistine 'development': British Railways consider a sense of history an expensive luxury: fund raisers must fight for appeal with a dozen other worthy causes: preservationists differ among themselves: locomotives find purchasers more readily than do stations: too little is being done, and it is being done too late.

CHAPTER TEN

'Fifties and 'Sixties

The one-inch ordnance survey sheet no 172 provides, as do many other sheets, a potted history of British railway engineering. Straight across it from Ashford towards Redhill runs the line from Dover; this was built in an age when any day might bring a new geological hazard and railways themselves were comparatively untried. But from Headcorn a single track is still marked as wandering south with a certain respect for contours but a complete disregard for directness. By comparison with Cubitt's trunk route this appears prentice technology, yet in fact it is more than forty years younger and was not untypical of the rural branches built in that half of the century when railway building had become a mature skill and railway operation was confidently expected to pay its 5 per cent.

The reasons for this paradox—this apparent collapse of standards which can also be seen as an injection of realism into railway economics—have been discussed earlier. Perhaps all that needs adding here is that as the balance of the eighteenth century and the dynamism of the early nineteenth slipped ever further into the past, as even the Great Exhibition became a historic memory and the queen herself faded into Scottish mists, the railways too were inevitably coloured with the more complacent extremes of *laissez-faire* philosophy. Thus, in earlier years governments had done something more than lay down safety standards: without presuming to plan the national system as a whole they had, for instance, chosen between rival schemes for railways to Brighton, Scotland and Ireland, and had even acquired a right of nationalisation. But after the midpoint of the century the *ad hoc* British system seemed to be working so well by comparison with its European counterparts that it became almost unthinkable to suggest that a measure of state intervention could ever be beneficial. The board-rooms became imperial divans, to be challenged by the body politic only when this smelt a danger of monopolisations

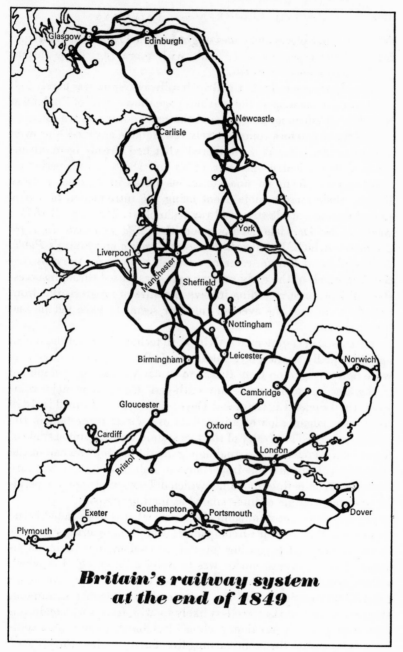

Britain's railway system at the end of 1849

Figure 9

which in other places and times might be called rationalisations; and branch lines especially were built with more regard for instant profits than long-term needs.

Even by the early 1850s the British railway system was barely half built. But, as the map on the previous page shows, most of England at least had an adequacy of lines, and the half-century which followed was hence to produce comparatively few notable advances—or even achievements, when it is considered what had already been accomplished. In particular, the later 1850s and 1860s (a dull period on many counts) form an unadventurous phase in national railway history, their most notable event being the introduction of a new material from outside strict railway technology. This, up till 1855 a work by G. Drysdale Dempsey entitled *The Practical Railway Engineer*—which, like Bree's *Railway Practice* and Simm's *Public Works of Great Britain*, remains a valuable source for the constructional methods of the early railway age—has passed through successive editions, each recording its own measure of progress. But from that date onwards the average engineer seems to have felt he had little to learn.

This blend of commercial activity with technical stagnation, which applied equally to the great works which British railway engineers were now engaged on from Russia to South America, is exemplified in the area just mentioned—the south-east. Here a new major company, the London Chatham and Dover railway, was formed in 1859 from an amalgamation of minor lines as a direct contestant to the South-Eastern. The rivalry of the two system for the limited trade of Kent, which ended in a virtual amalgamation before the end of the century, is of considerable importance in industrial history as a rare example of competing companies which did not simply operate parallel lines but became as intertangled as mating octopods.

From the present standpoint, however, what is to be noted is the way in which these Kentish companies lowered engineering standards in the interests of beggaring (or at least challenging) their neighbours. Thus, a typical device was to build a 'blocking' line which served virtually no purpose other than to secure a territory. Although the LC&D's engineer was the quite eminent Thomas Crampton, great structural works were less likely to arise from such conditions than were complex junctions such as Chislehurst. Even from a commercial viewpoint the profit-earning life of many of these lines proved brief; and today, in rural Kent as elsewhere, the undergrowth tangles across roadbeds which were the objects of bitter legislative battles a century and more ago.

Similar feuds affected the Brighton company, so that angry wrangles marked the building of, for example, a humble exurban branch to Caterham. But westward the land was a little brighter so far as engineering works were concerned. Down the boundary between the Brighton and the South-Western territory, for instance, Thomas Brassey had in 1858 completed at his own expense a 'direct' route to Portsmouth. (Curious as the procedure today sounds, it was not unique in this period for prosperous contractors to build major lines which they hoped to lease out to operating companies, another example being the London, Tilbury and Southend railway.) This Portsmouth railway incorporated gradients as steep as 1 in 80; but its profile was at least not so saw-edged as that of the LC&D's main line, and it involved the moving of over 1½ million cubic yards of spoil.

The next year the 'direct Portsmouth' was in fact leased to the LSWR. But the same company was simultaneously engaged in completing a scheme—which had been interrupted by the recession, to the disgust of the residents of Salisbury who did not benefit from a main-line service until 1857—to create its own access to the deep west. Again there were some heavy gradients; but again the standards aimed at were not unworthy of the pioneers. This line was opened through to Exeter in 1860.

Amid the Wessex chalk, too, Brunel had built for the GWR a substantial mixed-gauge branch to Weymouth, opened in 1857. Of more significance for the future was this company's steady formation of a southerly main line running from Reading via Newbury and Hungerford on to Westbury and Frome. A number of other lines, such as the joint Somerset and Dorset railway, also had a few engineering works to add to their charms. But the outstanding structural event in the west was the formation of the last link in the GWR's main route, the crossing of the Tamar outside Plymouth by the Royal Albert bridge.

This too can be viewed as a project left over from the 'mania' days. It was noted earlier that Brunel had planned a timber bridge at Saltash, that the Admiralty insisted on the high and wide clearances which implied a metal structure, and that the project hence ran into financial difficulties. In fact, work had begun on the key central pier before 1849. But it was abandoned, not to recommence until over five years had passed; and by that time Brunel had had the advantage of studying Stephenson's success with the Britannia bridge.

Now he proved correct the journalist who had commented in North Wales that 'anything so mighty of its kind had never been seen *before*: *again* it would assuredly be'. For Brunel decided to adopt the same practice of prefabricating long girders and then floating and jacking them into place, so avoiding the need for any scaffolding which would interrupt navigation. Being Brunel, though, he did not imitate Stephenson's box girders but instead found a more expressive way of overcoming the weaknesses of wrought iron. Ten years junior to Menai this work at Saltash survives, in use and almost unchanged, as one of the handful of outstanding monuments of British railway engineering.

A few years earlier Brunel had been faced with the problem of bridging the Usk at Chepstow on the South Wales railway. The work there presented several unusual features, among them being the facts that there was a cliff one side and a plain the other and that I. K. Brunel sunk his piers (as his father had sunk the shafts of the Thames tunnel) by building them up on sharp-edged cutting 'shoes' inside which excavation proceeded. Of greatest note, though, the form of truss he devised for Chepstow; for though this itself has recently (and, for once, necessarily) been demolished, it proved a predecessor for the twin girders used at Saltash.

The Chepstow span was some 300 ft long; those at Saltash were each of 465 ft, weighed over 1,500 tons, and were generally bolder, though economy restricted their width to that of a single track. But in each case the system employed was a variant on the suspension principle. The major member consisted of an iron tube, circular in section and bowed gently upwards at Chepstow and elliptical and more markedly curved at Saltash. Tied to this there hung a chain (which in the case of Saltash was largely salvaged from Brunel's own Clifton road bridge); and from it in turn there depended the rail-deck itself. The assembly had the suspension bridge's advantage of a balanced load without the tendency to vertical and lateral 'whippiness' which had led to the exile from railway practice of the pure suspension system.

It also had an aesthetic appeal of its own. Its principle, together with that of some other types of railway bridge used both before and afterwards, is outlined opposite.

The Royal Albert bridge (whose erection is described by L. T. C.

STATIONS—THE TUDOR
DESCENT

35 *Top* Gilsland, Cumberland
36 *Middle* Lairg, Sutherland
37 *Bottom* Battle, Sussex

VARIATIONS ON A THEME

38 *Top* Woburn, Bedfordshire
39 *Middle* Cromford,
   Derbyshire
40 *Bottom* Menai Bridge,
   Anglesey

Principal types of bridge structure

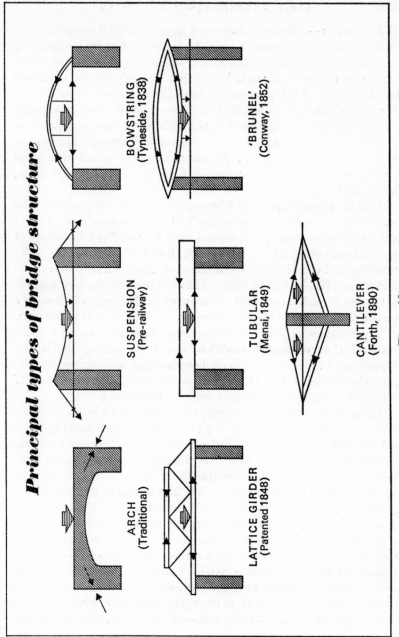

BOWSTRING
(Tyneside, 1838)

'BRUNEL'
(Conway, 1852)

SUSPENSION
(Pre-railway)

TUBULAR
(Menai, 1849)

CANTILEVER
(Forth, 1890)

ARCH
(Traditional)

LATTICE GIRDER
(Patented 1848)

Figure 10

Rolt in his life of Brunel) deserves a more detailed description than it can receive in the present book; for in addition to its engineering interest it ended the long isolation of the Duchy of Cornwall. What cannot be forgotten, though, is that in a dramatic fashion it formed the final work of the man whose name is still incised on its towers. For the difficulties at Saltash itself had been small by comparison with those which Brunel had faced in the building of his last great ship, and these had broken his health. He lived long enough to cross the bridge on its completion in 1859, but as he did so he had to be propped up on a specially-built invalid carriage. A few weeks later, Isambard Kingdom Brunel was dead.

Within months he was followed by Robert Stephenson, the only railway engineer to lie beside Telford in Westminster Abbey; and a few more months took away Joseph Locke. The triumvirate of giants, born within a few years of each other in the first decade of the century, died within a twelvemonth, worked out before their sixtieth birthdays. William Cubitt followed in 1861; and ironically it was left to Charles Vignoles, once half-bankrupt and half-disgraced, to make a second fortune and to die with a weight of years as well as honours.

These men who grew up with railways have been called the last heirs of the renaissance tradition; and their passing marks the beginning of the end, so far as civil engineering is concerned, for the individual pioneer and the signed masterpiece. But the work of railway building itself naturally went on unchecked.

In the early 1850s, for instance, there had been a boom in South Welsh coal which had led to a proliferation of lines serving those valleys which the tramways had penetrated a half-century before. The most notable work to come from this expansion was Thomas Kennard's Crumlin viaduct across the Ebbw Vale—a breathtaking if never beautiful iron structure, nearly 200 feet high and rightly regarded as one of the prides of the Principality, which was recently dismantled for scrap. A decade later, mid-to-north Wales acquired a main line as—aided by the strains of 'See the Conquering Hero Comes' played by a lordling on a *cornet aux pistons*—David ('The Ocean') Davies and Thomas Savin's Cambrian railways (*sic*) were opened from Welshpool to Aberdovey, and there branched to serve the coast from Aberystwyth to the tip of Merionethshire.

This enterprise was shortly followed by the Vale of Llangollen line, the Bala valley route, and an incursion from the LNWR which passed through Wales's longest tunnel to end amid the quarries of Blaenau Ffestiniog. But the main Cambrian line is of special interest for its 1 in 50 inland gradients, its benching along the rocky coasts, and the fact that more than a century afterwards it was still carrying traffic over several estuarine bridges of timber. The greatest of these, 800ft long and with an opening metal span, is at Barmouth: others further to the north carry toll roads and might hence survive the threat of the line joining its neighbours in closure.

In the English Midlands the main developments of this period were associated with the Midland railway itself. With its extension to Manchester complete (though later to be supplemented by a new route utilising the Edale valley slightly further north and passing through the Totley tunnel which, at over $3\frac{1}{2}$ miles, remains the longest mountain bore in Britain), this was now by far the most important of the companies lacking their own access to London. But the Midland was not at all happy with that distinction, and had already engaged in a steady expansion southward from Leicester to Bedford and beyond. As early as 1858 it had connected with the Great Northern at Hitchin so that its trains could run into King's Cross; but inevitably the terminus became overcrowded, and after the traffic chaos of the exhibition year of 1862 the Midland decided to lay its own metals into the capital.

As an engineering enterprise the route is remarkable mainly for its tight curves; but after squeezing through north London it ends in splendid style in W. H. Barlow's great train hall at St Pancras. Completed in 1868, the ridged roof of this is nearly 250 ft wide and 700 ft long and for many years covered the largest such area in the world. The ironwork (as founders' plates still announce on those pillars which are tied together in vaults below track level) was cast by the Butterley company, founded nearly a century before to supply Jessop and Outram's tramways and nearly a century later still busy in providing equipment for BR electrification schemes.

The St Pancras train shed was later to be paralleled by those at Manchester (Central) and Glasgow (St Enoch's), and was to set a pattern for New York's first Grand Central station too. In 1876 it was complemented by Gilbert Scott's magnificently romantic and eclectic Gothic hotel, a notable contrast both to the hall itself and to Cubitt's functional work at neighbouring King's Cross. For the St Pancras hotel, despite its charms, is *not* 'railway architecture'.

The year of 1861, which had seen the completion of the Midland's first trans-Pennine route, also witnessed a new expansion by an historic company which had long remained independent if somewhat stagnant, the Stockton and Darlington. This was now tempted westward by the attractiveness of exchanging ironstone for coal; but its route over the spine from Barnard Castle, Co: Durham, to Tebay in Westmorland was a rough and lonely one, calling for no fewer than twelve viaducts.

The general engineer here was Thomas Bouch—an eccentric genius who was later to be the subject of tragic attention. But the two metal viaducts at Glen Belah and Deepdale, each about 1,300 feet above sea level and nearly 200 feet higher than their valley floors, were designed by R. H. Bow, the mathematically-minded engineer whose 'notation' for stress-analysis in pinned structures is still taught. These great works of the last age of cast and wrought iron were outstanding of their type and of real technical and historic value: the piers of Glen Belah, in fact, may have represented the most ambitious use ever made of castings. They were perhaps comparable in importance to Eiffel's steel bridges in south-central France; but a few years back they met the fate of the Crumlin viaduct and were destroyed for scrap.

Other developments in the north included the joining of the Furness (1857) and Wirral (1863) lines to the national network, while on the Scottish border a new link was completed directly from Carlisle to Edinburgh—the Border Union railway of 1862. But for social as well as engineering reasons the most important railways of this period were those which penetrated, if initially only in the interests of tourism, into the depopulated spaces of the Highlands.

Here the *west* coast companies' route had been prolonged via Perth to Aberdeen by 1852, the engineer being the Joseph Locke who had brought this line all the long way north since his work in the English Midlands fifteen years before. It was left to a local man, Joseph Mitchell, to take over the direct route from Perth to Inverness, the 112-mile Highland railway which was completed in 1858 after only two years of work.

This line, beginning and ending near to sea level, crossed two watersheds at 1,052 ft and 1,484 ft—the latter being the highest point on any British main line. Considering the nature of the country and Mitchell's limited budget it was a considerable achievement to keep the ruling gradient down to 1 in 70, but long runs at this figure were inescapable. Such lines as the Highland, indeed, were for social and climatic as well as geological reasons far more reminiscent of overseas (for example, of Austrian) usage than of English standards; for they were frequently single-tracked and completely reversed one practice which the Stephenson generation had inherited from the tramway engineers. Whereas the latter had interrupted their long and almost level runs with short and steep gradients, Mitchell had to take steepness as his norm and use his skill to relieve it at strategic intervals with brief level stretches where steam locomotives could regain their breath.

Major engineering works on the Highland railway include the beautifully situated viaducts at Struan, Dunphail and Killiecrankie. Whether or not as a result of the influence of Locke, there are only two tunnels. Thanks to the backing of the railway-minded Duke of Sutherland the line was prolonged in 1874 for a further 160 difficult miles (twice the direct distance) to the extreme northern point of the British railway system at Thurso: this involved a detour on a spectacular traverse between Invershin and Lairg. A few years earlier a cross-country route had been opened to a ferryhead for Skye, and before the end of the century other typically Scottish branches were to be completed—via gradients of up to 1 in 45—to Oban and Ballachulish. It is noteworthy that the (English) west coast companies, having sponsored through lines in the *eastern* Highlands, then returned to the tangled Isles by way of long and lovely branches through the glens. Contrariwise, the east coast companies proper found a way north from Glasgow, crossing some fine viaducts as they went, to Fort William and beyond.

A number of miscellaneous works belonging to this period have been passed over in the above account, such as the tunnels—all more than two miles long—at Cowburn, Dinsley and Harecastle and the viaducts at Gothwood (Nottinghamshire), over the Tees at Yarm, at Hownes Gill in Durham (built by Bouch in the unusual material of firebrick), at Harringworth on the border of Rutland (somewhat later, but the longest combination of earthworks and arched viaduct in Britain), and—a forgotten work, this, dating from 1869 and later destroyed by storm—across the Solway Firth. Forming a group of their own, though, are the bridges which brought the south-of-the-Thames companies closer into the heart of London.

It has already been mentioned that the penetration towards city centres was so costly a process that companies had often had to treat their last mile or so as a luxury to be added after opening. The Birmingham line, for instance, had for some years had to delay its deep works near New Street, while in London the transpontine systems had for decades terminated at such remote points as Lewis Cubitt's 'West End terminus' of the Dover line, which was at Bricklayers' Arms in the Old Kent Road. But by the early 1860s the capital—and its rail-generated commuter traffic with the southern counties—was growing so fast that the rival companies were inspired to bring their lines across the river, especially as they could there link up with the underground system which was beginning to bind the northern terminals to each other and to the West End as well as the City. This purpose the circuitous North London railway had singularly failed to achieve.

So London River, which had never before been crossed by a railway, was, despite the severe financial crises of this period, bridged five times between 1863 and 1868. In all cases the form used was that of a series of latticed girders following the pattern of the 'Warren truss' which had been devised in 1848. The first such bridge was that of the West London railway, a short but important link between a number of different systems whose own history went back to 1844. There followed Hungerford or Charing Cross, Cannon Street, Grosvenor or Victoria (now rebuilt), and Blackfriars.

Almost without exception these bridges were designed by John Hawkshaw, a very versatile engineer who thirty years before had designed the first 'fixed' signals but perhaps a man of greater technical accomplishment than aesthetic sensibility. For even when seen through the most Whistler-misted eyes, these latticed spans form a drab series. The West London bridge at Battersea is perhaps the best, Hungerford is of interest in that its towers date from a suspension footbridge built by Brunel twenty years earlier (whose chains went to Clifton, to replace those which had themselves been cannibalised for use at Saltash), and some of the company insignia are worth attention. But it cannot be claimed that the thousand tons of cast iron which the City corporation insisted be added to Cannon Street bridge remind one of much more than paste diamonds hanging round a very scrawny neck.

Perhaps the lattice girder differs from almost every other mode of bridge construction in being inherently unlovely. This impression is certainly strengthened by comparing two bridges opened in the same year as witnessed the first Thames rail crossing, 1863. The latticed span at Runcorn in Cheshire, 300 ft in length, adds little to the wastelands of the Mersey; but at Buildwas in Shropshire John Fowler designed the 200ft Albert Edward arch of cast iron which, like its sister near Bewdley, remains an example of sophisticated design in a material which had served the tramways and railways nobly for a century but which was soon to be relegated to their sidelines.

For the 1860s were the years of the second—and sharpest—of the three great transitions brought about in the practice of railway civil engineering by the introduction of a new stress-bearing material. This material was mild steel, which eventually replaced iron just as iron had replaced timber. The key dates and data of this revolution, beginning with Henry Bessemer's patent of 1855 for the 'acid' converter and moving on to the various modifications of the open-hearth process associated with the names of Siemens, Martin and others, should be looked for in W. K. V. Gale's work in this series. All the present chapter can do is end with a note on the introduction of mild steel in railway practice.

Even by 1860 the railways of Britain represented a huge slice of the nation's involvement with civil as well as mechanical engineering; and just as Britain's major railway companies differed from their overseas counterparts in building their own locomotives, so they did in self-producing many of their other requirements. They had, for instance, large ironmaking plants and rolling mills of their own; and one of them, the Midland, saw the possibilities of mild steel rails as early as 1857.

In that year a section of track at Derby, subject to such heavy use that it had had to be remetalled with wrought iron rails every three months, was experimentally laid with mild steel rails rolled at Ebbw Vale from a billet cast in the Forest of Dean. These endured for over fifteen years, carrying 500 trains a day; and though this proved exceptional it was found that mild steel could be counted on to give about five times the life of wrought iron rails at an only slightly increased initial cost.

Mild steel, like wrought iron itself, was not introduced without development troubles after the carefully-made experimental samples gave place to mass-produced rails manufactured with less strict supervision, and some composites were tried. But by the late 1860s steel had become the accepted rail material, and at least one railway company had done something to repay its debt to the new science of metallurgy. For at Crewe the versatile John Ramsbottom had in 1864, after careful comparative tests, not only set up a converter plant (itself serviced by one of industry's first control laboratories), and even begun experimenting with steel sleepers, but had speeded up the reversing of his two-high stands by coupling them to their steam engines via a locomotive-type link gear. Rolling stands of this type produced the ribs for (for example) the St Pancras roof.

Mild steel also made an early appearance in a host of railway accessories, such as the latticed masts of signal posts and point control rods. The story of its introduction into general structural practice is a longer and more complex one, and is developed in other books in this series. It is generally true, though, that the authorities responsible for inspecting bridges and viaducts were more cautious in Britain than overseas, and that not until the late 'seventies could it be said that the age of cast and wrought iron was finally giving way to that of mild steel structures.

CHAPTER ELEVEN

Fin de Siècle

The concept of industrial archaeology is more than a century old, but only in recent years has the phrase itself become so fashionable as to become debased. In particular it is increasingly used as a synonym for technical history, whereas in reality there are two distinctions between these complementary disciplines. For any archaeology should be more concerned with the surviving artefacts of an age than with its written records, and in addition the very word implies the existence of a *terminus ad quem*, a date at which the archaic ceases to *be* archaic.

It might be over-scrupulous to insist that industrial archaeology should confine itself to events before the year A, for A will vary with the particular field of study. Thus, a terminus adopted for watermills would probably be earlier than that apposite to rubber technology, while one has yet to hear of the vocation of automobile archaeologist. But even this may come in another generation—which further suggests that the termini should themselves be allowed to crawl forwards in time. In the postwar years an 'historic' building has (in a British context) in fact come generally to mean one built before 1840 rather than 1800, while individual buildings dating from the 1930s are now scheduled for preservation. One would, however, have some sympathy with a pedant who queried whether industrial archaeology could pretend to being an academic discipline when it could not even define its subject-matter.

Some accepted guidelines are hence greatly needed. Meanwhile one must set one's own, and it seems reasonable to suggest that the *terminus ad quem* for the structural if not the tractive side of railway technology should be placed at not much less than a century ago. For in the 1870s (as the graph overleaf shows) the end of a long and, decade by decade, steady period of physical expansion was drawing in sight. Major new works still remained to be acheived and new

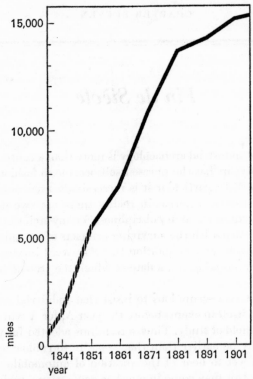

Figure 11 Route mileage of railways open in England and Wales at 'Census-year' (ie. 1851, 1861, etc.) intervals throughout the 19th century. This graph is intended as an extension to Fig. 8. It should be noted, though, that it omits the Scottish mileage

methods and materials to be mastered, but the methods and materials bore more relation to those of today than to those of the tramway age. And, above all, before Britain celebrated the golden anniversary of the opening of the Stockton and Darlington railway she was no longer *the* railway nation.

Up to and including the introduction of mild steel, this book has largely described British innovations which were also world ones. But by 1870 one European country after another was becoming able to develop its rail system without outside advice (even when these systems implied ten-mile mountain tunnels) and indeed to export its techniques, whilst in the USA the age of the civil war was giving way to that of the transcontinental railways. British engineers were still achieving great things in distant parts; but in their homeland the railway story for the last quarter of the century became of almost parochial importance.

There are hence good arguments for closing this book within a few more pages. But a story should not be left in mid-air, and there are better reasons in favour of carrying this account of railway civil engineering briefly forward to the present day.

The 1870s, then, witnessed their quota of long-distance railways, such as those built by the LSWR in Devon which produced works of the importance of the Meldon viaduct—a metal structure, and hence one now demolished. There were also substantial extensions in the industrial Pennines. But the only new line of the first importance was the final component in the Midland's expansion. By this, the Derby-based company became able to run over its own metals all the long way from London to Carlisle.

This last link—still open to passengers, though under threat of closure—branched off the existing route at Settle in Yorkshire and headed north-westward for seventy-two miles through an almost deserted countryside. It was built between 1869 and 1875, and is of further note in that it was the last great railway constructed by means which had barely changed since Robert Stephenson's time. Since 1842 mechanical excavators and other steam-powered equipment had been increasingly used on American railroad sites; but almost every attempt to apply them in Britain had ended with those words, still too familiar, that they were 'not suited to local conditions'.

Hence the Settle and Carlisle railway (which for political reasons had to take far from the easiest route) was built by men working almost unaided in a countryside of extreme cruelty. The line's first surveyor (and predecessor there of J. H. Crossley), John Sharland, was snowed up for three weeks near the 1,150ft peak of Ais Gill and later died of the effects: bogs such as Batty Moss proved so treacherous that wheelbarrows were swallowed up unless provided with foot-wide wheels: and smallpox took its toll in the encampments of the last of the old tradition of navvies. For twenty-two miles the line was rarely less steep than 1 in 100, but it still involved some very substantial works including the 2,630-yard Blea Moor tunnel and nineteen viaducts. The most notable of these, of twenty-four arches rising up to 156 feet above the valley, is at Ribblehead: but that at

Smardale proved so difficult to build that its construction took more than four years.

The Settle and Carlisle was the last of Britain's 'heroic' railways: far more typical of a new age was the extension of the suburban networks of London and other great centres and the construction of more lines within city limits. In one such case, at Liverpool, this movement was to lead to a curiosity now unfortunately vanished—Britain's only 'elevated' railway in the sense of one built throughout on a metal substructure. Despite its late date of 1893 this is mentioned here for two reasons—because it was in many ways anomalous (its engineer, Greathead, for instance chose wrought iron rather than mild steel for its girders), and because there was a serious proposal to serve inner London too with a network of overhead 'crystal ways'.

In the event the capital chose to meet the pressing problems of easing its internal transport by burrowing below the surface instead. The story of London's underground railways, now more than a century long, forms an engineering history in itself; but it is a story which has been very thoroughly told by Charles Lee and others, and is perhaps marginal to that of more conventional railway technology. It should be noted, though, that the first underground lines were conceived as extensions to surface ones and, thanks to the GWR's interest, were even laid to broad-gauge clearances.

This system (as the last chapter noted) was begun as early as 1863 after a decade of negotiations, the first section opened being that between Paddington and Moorgate which still preserves most faithfully the aura of those smoky days of steam working. The engineers were the distinguished team of John Fowler and Benjamin Baker. A little later a similar City-directed line was commenced in the Kensington area; but it was to be more than twenty years before the 'District' and 'Metropolitan' systems were finally linked near the Bank of England to form an 'inner circle', and by that time both companies had thrown out long surface-level feeders into suburbia and even exurbia.

The reason for this long delay was that the network was built, not by true tunnelling, but by 'cut and cover'. And though apparently a cheap method of construction, this demanded access from above and meant that the lines had for the most part either to follow the course of existing streets (such as the Marylebone Road) or to clear new ones like Victoria Street.

Subsurface developments were not confined to London, for in the late 1870s similar work was begun in two other great centres. These were Glasgow, where the crossing of the Clyde which had long defeated bridge builders was now achieved by not only a self-contained underground railway but by the main lines too, and Liverpool. In this latter city the tunnel to Birkenhead, with its steep approach gradients of 1 in 27, became the first example of a true deep-level 'tube' railway bored with a shield and lined with pre-fabricated cast-iron segments, though in London the Brunels' Wapping–Rotherhithe bore had now been converted to rail use and by 1870 Peter Barlow and his South African pupil James Greathead had, with an improved shield of their own pattern, drilled a small-calibre tunnel below the Thames near the Tower.

Back in the City of London, the year 1884 which witnessed the completion of the cut-and-cover system also saw the passage of a Subway Act for the construction of those deep-level tubes which were now virtually the only means of squeezing more railways into the confines of the capital—though for commercial reasons these too tended to follow the street layout. They were of standard rail gauge, but were built to a more limited loading gauge than surface railways (the typical tube bore being under 12 ft), employed unconventional trackwork, and are hence even more peripheral to the theme of this book than were their cut-and-cover predecessors.

Though mooted as early as the 1860s, such tubes had to await a technology largely devised by the two men mentioned above, Barlow and Greathead. These carried on a tradition founded by Marc Brunel by devising better shields for boring tunnels through soft soil and lining them as they progressed. In the present century similar techniques have been used to prepare tunnels for roads, water conduits, sewers and conventional railways as well as under-ground lines; and it is true to say that in the 1880s Britain not only pioneered the development of urban subways but established a supremacy which still obtains in the field of tunnelling through clay soils. For obvious reasons the country was less involved in advances in rock tunnelling, since by 1869—when high explosives were set to work on the Mont Cénis tunnel and showed a fivefold saving in time as compared to gunpowder—few mountain works remained to be achieved at home. From 1840 onwards, however, British as well as continental and American inventors had played a part in the improvement of rock-drilling and similar pneumatic equipment.

The first of the London tubes was opened in 1890. It ran from King William Street to Stockwell, and most of it is today incorporated into the world's longest tunnel, the seventeen subsurface miles of the Northern line. A Greathead shield of the type used is still preserved at Moorgate station. Long before this system approached its present extent, however, Acts had been passed for similar but improved deep-level lines such as that original 'twopenny tube', the Central London railway. These lines ran into their own engineering problems; but the hindrances were more often man-made ones such as sewers than natural hazards like the quicksands below Euston.

Tunnelling was so important a theme in the final quarter of the last century that one must now return to the main lines—and, in time, to 1879. In that year a new railway was built to serve the still-booming coalfields of South Wales by way of a bridge across the Severn from Sharpness, Gloucestershire: three-quarters of a mile in length, this remained a massive example of multiple bowstring girders until it was seriously damaged by a navigational accident in 1960. But as old Daniel Gooch of the GWR attended the opening banquet for the Severn bridge his mind was elsewhere; for his own company was engaged on an even more ambitious crossing of an estuary renowned for its fifty-foot tides and its treacherous, spring-ridden soil.

Back in 1872 the GWR had at last been granted an Act for the building of a tunnel below these waters—a tunnel which would save an hour on the journey from Paddington to South Wales. Work did not begin until five years later; but with the aid of the manually-worked shields typical of the period it progressed fairly well, and in the approach cuttings a rare use was made of steam excavators. The two headings were hence within little more than a hundred yards of meeting when—on the day before the opening of the upstream bridge—fresh water broke in.

The tunnel was now entrusted to John Hawkshaw, whose first and very demanding task was to tidy up the flooded chaos: the hero here was one Diver Lambert, who on the second of his perilous underwater walks through the dark to seal a vital floodgate made use of the first type of 'scuba' equipment. In 1881 construction began again; but it was repeatedly interrupted by inroads of both fresh and salt water, and not until 1886 did trains regularly pass through the Severn tunnel.

Since then the tunnel has needed continual upkeep, ventilation and pumping out: until recent years some of the original machinery was in use for these purposes, and examples of it have been preserved. Nearly 4½ miles long (though with only two of these under water), it remains the longest such bore in the world. But it would soon have lost that title had not another great scheme of this age of tunnelling been defeated by military shortsightedness.

This was the Channel tunnel, a work first conceived in 1836 and energetically promoted from 1872 onwards by Edward Watkin. As a boy of eleven, Watkin had watched the opening of the Liverpool and Manchester railway: now, more than forty years on and by way of successful investments in the Trent Valley, the Manchester, Sheffield and Lincolnshire, the South-Eastern and the London Metropolitan railways, he had become the most powerful and piratical railway entrepreneur of the second half of the century and the equivalent of George Hudson. Not surprisingly, the leading railway capitalist of his age had called on the services of its leading railway engineer, John Hawkshaw. But their efforts were wasted for reasons which had little to do with feasibility, the scheme which would have been the greatest of all memorials to Victorian engineering was vetoed by the queen herself in 1882, and today, when the 'Chunnel' is needed as never before, there remains only a mile of exploratory heading leading offshore from the Shakespeare cliffs.

Despite this accent on tunnelling, there were water-gaps left in the last quarter of the century for which traditional bridging remained the better solution. In particular, as a glance at a map will show, the journey northward from Edinburgh to Dundee and thence Aberdeen implied either a thirty-mile detour inland round the firths of Forth and Tay or the use of boats which became renowned for their discomforts. In 1850 the North British Company, which served this coast and was very conscious of the more direct services available from the west, had introduced the world's first train ferries in order to ease the agonies of the journey. But these were hardly a permanent solution.

The estuary of the Tay south of Dundee, though wider than that of the Forth north of Edinburgh, was shallower and less used by shipping; and so it was here that work on a bridge was completed in

1878. The engineer was the Thomas Bouch who had distinguished himself on the Stockton and Darlington extension, and his plan of a series of lattice girders supported on more than fifty piers across the two-mile-wide estuary was in essence a conventional one for the times. But Bouch had always cut his margins very fine, and in this case errors of design were compounded by gross negligence on the part of the contractors and inspectors. In 1879, only a year after its opening, the Tay bridge collapsed under load in a storm and the most famous accident in the history of railway engineering. Seventy-five lives were lost, the great McGonagall wrote his fearful ode, and to these disasters can be added the death of Thomas Bouch in national disgrace only a little later. As at Saltash, though in a different sense, a great bridge was also an engineer's tombstone.

But the crossing of the Tay had already proved its worth, and eight years later a stouter if less distinguished structure, which made some use of the materials of its predecessor and which hence became the last of the world's great structures to be based on wrought iron, was opened to the designs of W. H. Barlow (the same as had designed rails for Brunel nearly half a century before) and his son. This remains today Britain's longest rail bridge, and indeed it is only in very recent years that such Victorian mammoths have been challenged by road bridges over the Tamar, the Forth and the Tay itself.

'Mammoth' is a word which can be applied in more than length to the Forth railway bridge, which was completed in 1890 and was opened (as a portent of a new century) by the future Edward VII. Here the busy firth was 200 feet deep and Bouch had planned a suspension bridge, with two spans separated by a central tower on the island of Inch Garvie, which would almost certainly have proved a death-trap. The Tay tragedy hence had at least two beneficial side-effects—a more scientific study of the effects of wind pressure in general, and a rethinking of the design for the Forth bridge in particular.

The solution presented—by the Benjamin Baker of underground railway fame and Henry Fowler—was to build three great structures, of the type mistermed 'cantilevers' (which, as Fig 10 shows, were actually diamond-trusses), to be linked by 'floating' girders: each of the main spans was 1,700 feet long, or nearly four times the length of Menai or Saltash, and was some 150 feet above water-level.

THE DEVELOPMENT
OF THE METAL BRIDGE

41 *Top* Crumlin, South Wales
42 *Bottom* The Forth Bridge

ROMANCE
AND
FUNCTION

43, 44 St Pancras,
London

The Forth bridge (which, like its predecessors, is described in more detail in a companion book) was the first major British work to be fabricated from mild steel. Thanks to its system of construction its outline is as characteristic as that of Saltash: whether it is as aesthetically satisfying as it is impressive is more debatable.

Another, smaller, later and less-famed Scottish cantilever bridge is, however, of much neater design—as might be expected from the fact that one of its engineers was I. K. Brunel's son Henry. This is situated at Connel Ferry, between two tidal lochs near Ballachulish and the Falls of Lora. Five hundred feet in span, it carries also a toll road and so has survived the closure of the railway itself.

The Connel Ferry bridge was not opened until 1903; and since this chapter has thus trespassed into a new century a group of works may be completed by a mention of the King Edward VII bridge at Newcastle. This plain structure, designed by Charles Harrison to relieve the traffic congestion and need for train reversal caused by there being no by-pass to Stephenson's old 'high level' bridge, was completed in 1906.

All these tunnels and bridges suggest a still-expanding system, and as a measure of railway *utilisation* it is worth noting that *Bradshaw* contained 98 pages in 1845, 270 in 1864, 422 in 1884 and its peak of 1,270 in 1914. It was, however, a system which was now beginning to feel the first blasts of competition; and these came, not from the open road and the internal combustion engine, but from the railway's own child—the urban and interurban electric tram, with its low capital costs and close-to-the-door service. As John Snell's book mentions, the 1880s also witnessed the railways' own first experiments with electric traction, though it was not until the new century that this was introduced on a scale which involved substantial engineering works.

Meanwhile, after nearly half a century of steady expansion, the graph of growth was flattening off as Britain approached a peak of around 20,000 route miles of railway. Whether this length was more than should *ever* have been built is an open question, since the Victorians can hardly be reproached for not looking forwards to the era of the 40-ton lorry and the Jumbo jet by an age which cannot even keep up with its own transport demands. Perhaps some of the claims of late-nineteenth-century promoters that their latest rural branch or newest inter-city line triplicating existing services was performing a much-needed public service were disingenuous. But (as Dr Johnson knew) men are seldom so harmlessly and even usefully employed as when out for a quick buck; and from the cut-throat economic chaos of the great railway-building years between 1825 and 1875 Britain had gained a transport system more than able to sustain any demands which not only that period but a further seventy-five years could lay upon it. The Victorian railway system may in retrospect appear wasteful; but it often proved a great deal more serviceable than the planned networks of more logical nations.

Even after the 1880s this growth was not fully arrested. It should be remembered, for instance, that the figures quoted in this book generally refer to *route* miles; but the last two decades of the century were characterised by a great deal of work on line quadrupling (particularly within a radius of some sixty miles of London) which substantially increased the national total of *track* miles. In several locations—for instance, north of Bedford and in the Redhill, Surrey, area—these new lines occupied sites quite independent of the original ones.

Finally two main line companies achieved, though for very different reasons, a fresh impetus at the close of the century. The first of these was the GWR, which in 1877 had laid its last broad-gauge track. Thereafter the 7 ft gauge became even more manifestly a lost cause; and finally, over a single weekend which witnessed perhaps the most intense work of railway engineering ever performed, all the surviving broad-gauge metals were replaced by conventional ones. The last broad-gauge train ran on 21 May 1892.

Lovers of the railway past, even then numerous, naturally wept at the change or addressed commemorative odes:

> . . . Your place,
> Knows you no more; a pygmy race
> Usurps the glory of the road,
> And trails along a lesser load

wrote one poet to a broad-gauge locomotive. But now that it had emerged from its backwater the GWR could turn its attention to much-needed improvements in its routes as well as its motive power. The detours involved in its way to Exeter via Bristol as well as to Birmingham via Oxford had given the company the sobriquet of 'the Great Way Round'; but in the closing years of the last century and the first of the present one nearly a hundred miles of new cut-off routes were opened, the most important being those which passed through Bicester (Oxfordshire), Badminton (Gloucestershire) and Somerton (Somerset). The last of these involved a difficult crossing of the Parett marshes, the foundations of the roadbed being reached fifty feet below their surface.

Furthermore, the individuality of the GWR survived not only this physical change but later administrative ones: indeed, for all the etiolating efforts of BR it is not even yet quite dead. By contrast, the new trunk railway which was built in the 1890s has now virtually vanished from the map of Britain.

The creator of this was—perhaps inevitably—the Edward Watkin who has already been encountered as a Chunnel enthusiast. Reference back will show that his empire comprised two main parts—the old MS&L system, and a pair of lines in the home counties which were physically linked in south London, the Metropolitan and the South-Eastern. Between them, the latter provided a Watkin-controlled route from the Chilterns to the Channel; but the ambitions of the last tycoon of the railway age did not end at Dover, and in addition to his Channel tunnel activities Watkin held interests in the French Chemin de Fer du Nord. He could thus envisage a through service under his aegis all the way from Manchester to Paris, except for a gap between south Yorkshire and the Chilterns. This gap he now resolved to fill with a new line which should absorb the MS&L under the name of the Great Central railway.

So much background is necessary to appreciate how a main line came to be built so late and (to a large degree) in such an unpromising situation. The extension of the MS&L south to Nottingham, through two tunnels to Leicester, and on to Rugby could perhaps be economically justified, as could the corresponding prolongation of the 'Met' to Aylesbury—though this latter gave the London underground map a suspiciously lopsided look. But the ninety-mile crossing of the Midland plain by way of Brackley, Buckinghamshire, took the GC through country with less economic potential than now lay about the competing routes to the north.

Extra expense was incurred at the London end when it was decided to build a new main-line terminal near Baker Street, for cricket-lovers insisted that a tunnel be cut under the threatened Lord's ground. (This seems the only occasion where a railway was ever realigned for sporting reasons, though Brunel had had to consider the rights of Dorset archaeologists.) Hence Marylebone itself emerged as the humblest of London termini, though faced with a very fancy hotel. Even so, the GC—which was completed in 1899, two years before the death of its begetter—never paid a dividend; and less than seventy years later its central section achieved the sad distinction of being the first British main line to be totally closed and demetalled.

Economically it may have been a megalomaniac folly; but from an engineering standpoint the extended line surely deserved better of fate, as well as meriting the amount of space which it has been given here. For it was aligned to standards which had been forgotten for half a century, with a ruling gradient of 1 in 175, easy curves and a complete absence of level crossing; and this perfectionism was made possible by an extensive (on such a scale) pioneering use of powered equipment, including forty steam excavators. In anticipation of the Chunnel and an international trade which never materialised Watkin also laid out his new works to the generous continental loading-gauge, and even his stations were designed to allow for easy expansion. The Great Central, in fact, was arguably the all-round finest of all Britain's railways.

A tiny outlier of the GC—though in fact built as a private railway in the strictest sense—was the Brill, Buckinghamshire, tramway. This was typical of another theme of the closing years of the century, the comparative proliferation of 'light' railways or rural tramways. These were built in two waves, following on Acts passed in 1871 and 1896 which relaxed certain restrictions on construction and operation; but in fact these came too late to be very effective, for Britain was almost saturated with ordinary branch lines and there was no room for such national networks of narrow-gauge or unfenced ways as characterised certain continental nations before the first world war. Certainly (and this added to their charm) there was little in common between the amphibious tramway on Brighton front, the roadside Wantage, Berkshire, steam tram, and the mountainous Tal-y-Llyn railway in Wales.

Two such systems are worth a special, if brief, note in a book dealing with the structural side of railways. One is the Lynton and Barnstaple (1898), which crossed a 1,000ft crest of Exmoor and proved that even in England there was a place for narrow-gauge metals: its closure before the present age of preservation societies was a lamentable loss. The other example, opened in 1896 and still profitably active behind steam, is the Snowdon Mountain railway which climbs nearly 3,000 feet in 4½ miles on the Abt rack system. Its engineering, though striking, represents something outside any British tradition.

As the century closed a much more important innovation was making its presence felt. And though this was not a direct import from the Continent, it is true that Britain played no very distinguished part in the development for load-bearing structures of the material which (though known in Roman times) was to characterise the third and last major revolution in railway civil engineering—concrete. This is the more regrettable since in 1900 (and, despite the spread of tarmac, up to 1914) 'railways' and 'engineering' remained almost synonymous on the civil as well as the mechanical front; and at least part of the blame may be attributable to the fact that ever since the main-line network had taken shape in the middle of the century the railway civil engineer had become increasingly subordinate to his mechanical colleague.

The earliest British uses of concrete in railway (and, possibly, in any) practice appear to date back to 1867, when in west London the District line employed the material in a pioneering arched bridge near Gloucester Road designed by Henry Fowler. Four years later its depot at Lillie Bridge became the world's largest building to make extensive use of concrete. Not until the early 1890s, however, did concrete support the load of main-line trains: it was then used by the LSWR in several now-abandoned viaducts on extension lines in the west country such as Holsworthy near Bude, though only to form the piers which supported a steel deck.

The first all-concrete railway structures bearing heavy, dynamic and not directly compressive stresses appeared only when, at the very close of the century, the West Highland branch in Scotland was extended from Fort William to Mallaig. Neither bricks nor stone could be found economically in the area; and so this lovely branch (still, if contestedly, open) was carried on several bridges of concrete—the longest, 127 ft in span, being the greatest of its age—and on one curving, 21-arched viaduct of the same material at the head of Loch Shiel near Glenfinnan, Inverness-shire.

Yet even here the concrete was simply massed by Robert Mac-Alpine into forms which would have been familiar to the Stephensons, to the designer of the Causey arch, or to the Romans. At the debut of a new century there was none of that appreciation of the peculiar possibilities of concrete towards which the Swiss, French and Scandinavians were groping their way. In 1900 exciting new things remained to be done with the steam locomotive; but the shoulders of the British civil engineer seemed bowed below the weight of three quarters of a century of tradition. He had lost his sense of adventure; and he was not to recover it for another fifty years.

CHAPTER TWELVE

Past, Present, Future

If the previous chapter provided only a summary, this final one can offer little more than a footnote covering the story of railway engineering in the twentieth century before it looks at some prospects for the twenty-first.

As with other aspects of modern history, that story divides inevitably into three periods. Up till 1914 there is little enough to record, for Britain's Edwardian railway engineers showed even less interest in new techniques than had the late Victorians. The national rail network was virtually complete: there were as yet few signs of any need for it to be reduced: and it was in good shape physically. All that was demanded of the men of that long summer was hence to patch up a station here or a viaduct there.

The calm—and even complacency—of the Edwardian era is familiar ground: so too is the fact that the first world war proved doubly traumatic to Britain's railways. (According to Michael Robbins, indeed, 1914 marks the end of 'the railway age'.) Structures were then loaded as never before—and survived the ordeal, thanks to the Victorian way of building generously and building in duplicate. But the war also generated great advances in road and air transport, and to this new competition the railways found no answer. The Edwardian lack of enterprise may have accentuated their troubles, but at root the position was that, a century after they had begun to put the canals and stage coaches out of business, the railways' own technical monopoly was ending.

In 1924 a hundred or more independent companies—of which perhaps a score could be classed as 'major'—were amalgamated into the four great empires of LMS, LNER, GWR and SR. Together these badgered the governments of the late 1920s and the 1930s— governments with more immediate problems, in that age of bitter industrial unrest—for a 'square deal'. Their argument, still offered today and still unanswered, was that the admitted convenience of road transport was not paying its full social cost to the nation. For now, for the first time in its history, the British railway system was beginning to lose money.

In an effort to retrench, the first few hundred miles of very lightly used branches were closed, and one or two suburban extensions were axed. More constructively the 1930s were an age of belated experiment with various means of traction, though only one company plunged into widespread electrification. This was the Southern, which with its densely-used short-haul routes was in any case unique. Since the system adopted employed a third rail (rather than overhead wire) as a conductor, the changeover involved no substantial engineering works. But the SR was also active in modernising its ports and in rebuilding a number of stations.

This movement had begun with the completion of a replacement for the 1848 terminus at Waterloo. Work on this had started before the war, but it was not until just before the demise of the LSWR that it could be seen that the new station, though well planned from an operating viewpoint, was architecturally a monument of monstrous grandiloquence. The smaller SR stations (whether provincial terminals like Hastings or commuter facilities such as West Croydon) were rarely more elegant in their assembling of concrete blocks; but they at least avoided the peculiar jazz-age nastiness of a few new buildings erected by the LMS (and one or two by the GWR as well) during the 1930s. Good—or even interesting—railway architecture would seem to have passed fifty years before into the hands of the Americans and Germans (though *not* the French and Italians) were it not that Britain's one railway enterprise then operating as a public trust, the London Passenger Transport Board, hired a genius named Frank Pick to supervise not merely the design of new stations such as Arnos Grove but the introduction of an overall 'house style'. Pick's influence is still very active over this railway system which must (if in the face of economics) continue to expand if the capital is not to grind to a halt, but only after nearly thirty years did it begin to affect the main lines too.

In railway terms the second world war was very largely a repeat of its predecessor, though it led to an even greater backlog of maintenance work; and again a New Year's Day shortly after the coming of peace witnessed a total reformation of Britain's railways. The outcome this time was a long-debated but now total nationalisation.

It may as yet be too early to compile a balance-sheet of the last two decades, but there are clearly entries to be made on both sides of the ledger. On the debit sheet, the Britain of a car-owning age suddenly appeared over-railwayed to a degree which surprised even the most pessimistic, so that the line closures which began soon after nationalisation accelerated to an average of some 300 miles a year. The general principle was that of 'last in, first out', and hence an animated-film map showing Britain's railways expanding between (say) 1861 and 1881 would need only to be run back at the same rate to present a not-too-inaccurate picture of their contraction from 1951 to 1971—by which year the total route mileage had been reduced to under 12,000. Towards the end of the period this rate of closure naturally slowed; but it has not yet halted, and as this book goes to press the future of a number of semi-main and suburban lines (for the rural branch has now all but vanished) remains in doubt. According to prognosticators more level-headed than those who want to see a railwayless Britain, the nation's most viable mileage might be as low as the 1852 figure of 7,000—which is itself more than double the length of what Lord Beeching regarded as indispensable arteries.

This landslide of closures—which was also largely responsible for the unexpectedly swift death of the steam locomotive—mocked those who believed that nationalisation would at least make possible a coherent transport policy. In fact, the largest item to be written in red on the ledger is the total failure of national planners (whether at BRB headquarters or at cabinet level) to foresee, and either control or adjust to, social and economic changes. This failure is certainly not confined to the transport scene; but it shows there as vividly as anywhere. Snippets of motorway (less than 200 miles in all, but costing almost as much to build as the 6,000 miles of destroyed railway would today) have been constructed but not linked up: a new rail spur to Heath Row airport has been actively on the drawing board for well over five years: we have acquired part of a new tube railway but the Channel tunnel remains a dream. Organisationally there has been a frenzy of establishing, centralising, decentralising and recentralising, of commissions, boards and executives, with even the railways' liveries and house styles seeming to change at almost annual intervals in a frantic search for new images. But our boastfully affluent society, more than half of whose wealth now passes through the hands of central and local governments, has in the last two decades done less to improve its public transport than was achieved in perhaps five years of chaotic Victorian *laissez-faire*.

This is a heavy indictment, and not all the blame can be laid on a *zeitgeist* which has proved particularly unsympathetic to public transport. But for all their slovenly management and failures in foresight, British Railways *do* have substantial achievements to show as a result of the work carried out after the years of catching up with wartime neglect gave way to a period of active if much-interrupted modernisation. And these achievements are as numerous on the civil engineering side as anywhere.

There has, for instance, been a vast if unspectacular programme of track relaying; and this has implied not routine maintenance so much as the complete rebuilding of roadbeds in the interest of smooth and safe high-speed running. It has already been mentioned that BR's present rails—weighing 109 lb/yd—are of the flat-bottomed Vignoles form which experiments in the late 1950s showed to be more economical than the traditional British bull-head form in an age of rising labour costs. These are clipped to the concrete sleepers which are today cheaper than timber ones. Entire prefabricated sections are generally laid, with the rail-ends being butted together to form the continuously-welded tracks themselves made possible by the greater rigidity of concrete sleepers; and not only relaying operations but such routines of maintenance as ballast cleaning are increasingly carried out by machines travelling over the rails.

More noticeably, postwar examples can be found of almost every typical form of railway structure: these range from the new bores at Woodhead (the first major tunnel to be built in the present century) and Potters Bar (which marked a further advance in segmental techniques) to many dozens of bridges and short viaducts, often built in connection with motorway schemes. The rail flyover at Bletchley—which, typically, was part of a 'rationalisation' scheme whose viability became very questionable only a few years later—is of some historic interest as showing an early example of the railways' adjusting to the age of reinforced concrete: the result resembles the experiments of the late 'thirties almost as much as the elegant and assured, if repetitive, bridges of today.

In virtually every type of structure, from tunnel-lining to culvert and from cable-conduit to the windowless, electronic-box type of signal cabin which emerged from *another* reversal of policy, reinforced concrete in its various forms now dominates railway building as completely as it does so many comparable technologies: it is cheap, it is flexible, it is occasionally (and almost accidentally) graceful, it calls for little original thinking, and if it is a deadly monotonous material which also weathers most unattractively the present age has little thought to spare for such considerations. The more credit is due, then, to British Railways in that real imagination has entered into the design of at least the best of their 250 or more rebuilt stations.

Only a few such buildings—as at Durham—have incorporated local materials; and there have certainly been too many soulless tower blocks, such as Harrogate and Plymouth, which fail to announce themselves as stations and simply underline the functionalist heresy. But by contrast there are a host of modern British stations, large and small, which are as satisfying as any except the finest Italian examples and which appear far more distinguished than does the nation's recent architecture as a whole. A list of the best—in order of increasing size—would include some of the Southern Region's prefabricated suburban stations: Oxford Road (Manchester), Barking, and Bishop's Stortford and its neighbours in Hertfordshire: and Leeds (Central)— which will perhaps prove the final station to be built with an overall roof—and the admitted showpiece of Coventry. Even the new Euston in London, for all its faults, has features almost good enough to help one forget how much has been lost here.

Most of the works mentioned above are *ad hoc* ones, built in answer to special local needs of obsolescence or reroutings. There are, however, themes such as the increasing importance of cross-Channel ferry traffic which have been responsible for entire ranges of new railway building. And by far the most important of such themes has been the introduction of overhead-wire electrification.

Even in this field there were several visions and revisions, with Britain apparently lagging behind continental practice and initiative. Thus, the electrification of the Manchester–Woodhead–Sheffield line, though its civil works marked a great advance in the use of standardised concrete members, proved a dead-end electrically. But eventually a decision was made in favour of the 25Kv AC system, which was tried out on the Eastern Region (where it implied some fairly substantial general engineering) and was then applied to the great arterial network of the Midlands which fanned out from London to Birmingham, Manchester and Liverpool. The area involved was so extensive that the first stage of the project was a complete remapping of it by aerial survey.

The extent to which Robert Stephenson's structures survived even this radical reshaping has already been noted; but it would be unfair to his successors of the 1960s to pretend that it was *not* radical. Not only did the great majority of overbridges have to be raised (at one stage at the rate of one every day) to accommodate the overhead electric gear, but numerous underbridges were strengthened, track layouts were improved, and over fifty stations were rebuilt: these ranged from prefabricated halts to such major works as Birmingham, New Street, and Manchester, Piccadilly. All this, furthermore, went ahead with no serious dislocation of normal services, so that not just once but time and again there took place operations whose like, in their use of imaginative engineering techniques backed by sound logistics, had not been seen since that weekend nearly seventy-five years before when the GWR was narrowed. That despite modern machinery this work cost (in real terms) about as much as the 460 route-miles had needed to construct originally illustrates the difficulties of performing surgery on a living railway.

In the case of the LM electrification scheme, a change in the system of *traction* became the occasion for an immense volume of civil engineering and even architectural work only indirectly linked to the task of erecting standards and gantries, of building substations, and so forth. The interrelation between different facets of railway technique is indeed closer today than it was in Victorian times, so that a new telecommunications system (for instance) may make it possible for a formerly double-tracked line to be 'singled' and so be kept economically viable. But combined with this underlying unity is the fact that whole new technologies have come to maturity in recent years to make the work of today's railway engineer more exact, but also vastly more complex and compartmentalised, than that of his predecessor.

Thus even traditional railway arts such as bricklaying and protective painting have now become the subject of a scientific approach which extends from an examination of first principles through to the development of tools to help defeat one of the railways' most insidious enemies, the rising cost (and falling availability) of the artisan skills with which they were built. And on a more fundamental level the whole new subscience of soil mechanics has matured largely from the work of—and certainly to the advantage of—the staffs of Britain's railways. Today a clay embankment can be designed, or redesigned, almost as precisely as can a concrete viaduct.

If one attitude above all others distinguishes the modern engineer from his Victorian counterpart it is his reliance on systematic rather than empirical knowledge. A number of instances mentioned in this book illustrate the railways showing an appreciation of the scientific approach in advance of the standards of their times: for example, George Stephenson and Nicholas Wood's dynamometer experiments (which themselves followed the work of Telford and Smeaton), Robert Stephenson's approach at Menai, and the development of mild steel which at Crewe led to the appointment of staff analysts equipped with spectrographic equipment as early as 1864. But, with the exception of some work on locomotive design which was itself less than the age demanded, the railways' interest in applied science largely fell into that trough in which so much British enterprise was lost between late-Victorian times and the close of the 1930s.

Perhaps the most purposive of all developments in the postwar, British Railways era has been the establishment of an outstanding research department: with all respect to the contributions of France, Holland, Austria and (in particular) Japan, the new laboratories at Derby now form the deepest think-tank of railway technology in the world. On the civil engineering side alone this houses some of the most powerful equipment for fluctuating-load testing ever built. But civil engineering is not seen in isolation. It is rather one component of that complex which includes traction, signalling, traffic-handling and the allied crafts which have always been combined in the idea of a railway but which are more closely interlinked today than ever before.

Even the most traditionally-minded of BR engineers now recognises that the *possibilities* for a national railway structure viable in the year 2000 (and after) are being laid bare on the test-beds of Derby, rather than being left casually to emerge as of old from the hit-and-miss of operational experience. To indulge in any detailed speculation as to which of the possibilities will become realities would be to end this book with a digression: too much hangs upon the wisdom of governments, the fluctuations of economics, and the degree to which the nation can regain its early-Victorian energy and sense of direction. But in the last few years there seems to have been a certain crystallisation of opinion as to the probable overall shape of railway development in Britain in the last quarter of our tormented century.

The argument proceeds by three stages, of which the first is that, with an increasing population, there will remain a demand at least as great as the present one for some form of physically disciplined land transport. This phrase could include a system as remote from the Stephensonian railway as a road, equipped with linear-motor elements and with speed control centrally enforced, into which would be 'slotted' hover-cars, -lorries and -buses powered by fuel cells or small nuclear reactors: exaggeratedly futuristic though this vision may appear, it illustrates some elements of the thinking of those who now believe that the essence of a railway lies more in *control* and *guidance* than in two metal bars. The second stage of the argument, though, suggests that there is as yet no need to abandon those metal bars for any system of suspended monorails or concrete troughs, useful though these might be for supplementary transport in special conditions. For a few years back the city of Manchester commissioned an independent study to determine the ideal method of serving its commuter needs, and the answer came out firmly in favour of what was quaintly termed a 'duorail'.

Even duorails (which, by those who do not hold diplomas in town planning, are called railways) can come in several forms, however; and until recently it was suspected that the pattern for the next generation of main-line practice had been set by Japan's magnificent Tokaido line—a completely new railway designed in such matters as its block intervals and superelevation, as well as its very relaxed curves, for speeds of up to 150 mph. For though it is not long since one of the more egregious chairmen of the BRB issued a bull to the effect that his public was 'not interested in high speeds', the present railway thinking in Britain and similar countries is that one field in which that (comparatively small) balance of traffic which can turn loss to profit can now be captured is the custom of the expense-account traveller—and that, in any case, higher speeds imply a better all-round use of equipment.

In the attempt to push even peak speeds above the 100-mph mark technical problems become closely interrelated. But (and this is the third step of the argument) BR have now decided *against* building Tokaido-type duplicate tracks and *for* developing specially suspended rolling stock to run over existing metals. Such an 'advanced passenger train' is already promised to enter service a few years hence with 150-mph schedules; and though the French railways showed nearly a decade back that over 200 mph was possible in ideal circumstances with conventional stock on conventional tracks, a mere 150 mph promises a faster run from the centre of London to the centre of cities nearer than Edinburgh (or, with the Chunnel, Paris) than can any of the various forms of air transport.

For the future to be altogether hopeful in even the passenger department (and for freight the railways can hope to do little more than hold their own pending governments less dominated by the lobbyists of the lorry), one would need greater assurance that the railways will also be able to compete with the road for the custom of the holidaymaking family to whom pounds are more important than minutes. But this speculation is not wholly relevant here, whereas two facts are. One of these is that speeds of 150 mph will be possible without immensely costly rebuilding thanks only to the sweeping curves (even more than the easy gradients) adopted by Stephenson, Brunel and their contemporaries. The other is that, even should these looks forward prove too conservative and Britain after all shortly need some entirely new system of 'transportways', these will still have to cross valleys, pass through mountains and follow all the disciplines, today used in motorway practice, which were worked out by those pioneers who created not just a device but a technique, an architecture, an economy and a whole age.

Meanwhile the white-coated young technocrats of Derby, armed with their slide rules and with a knowhow and even a faith as great as can be found in any comparable technology, may seem very far removed from George Stephenson stumping out the muddy miles between Stockton and Darlington with his homemade theodolite under his arm, his son beside him to act as chain-boy, and a pint of claret and some shrewd questioning from Edward Pease awaiting in an inn at the end of the day. But as a late-night train has slowed past the eerily lit clerestory of the research establishment's great test-hall, one railway-lover at least has sometimes wondered if two grey Northumbrian ghosts—twenty-five years on now from Stockton—may not be stealing past the pulsing fatigue machines and across the cloister to slip through the analogue computer Mr Fairbairn's latest modifications to the grand design of Menai.

Gazetteer

There are well over a dozen museums in Great Britain with exhibits of interest to the student of railways: these fall into the three main categories of the specialist collections of (at least formerly) the BRB itself, the national, civic or private museums and departments devoted to transport and technology, and the more general museums containing a few railway exhibits. Not surprisingly, though, the majority of the larger exhibits in all these consist of rolling-stock and locomotives: apart from models, plans and illustrations, civil engineering is represented mainly by a few sections of unusual small bridges and by the early rail specimens which are particularly widely scattered. A somewhat fuller guide to railway museums (and the future of the major ones is currently in doubt) will hence be found in John Snell's book.

In his *Railway Relics* (Ian Allan, 1969) the present author attempted to compile a list—with dates of building and keys to location based on the Ordnance Survey—of 500-odd of the most important *in situ* survivors from the tramway and Victorian-railway eras. The book itself, which was intended as a *catalogue raisonné* of all aspects of the railway past, is mentioned here with some hesitation since certain errors regrettably slipped into its text; but its gazetteer is believed to be the only (and hence, without immodesty, the best) compendium of its kind available. The following notes are more selective: recapitulating and summarising information on a few dozen of the more outstanding works mentioned in the present book, they are intended mainly as a first guide to those relics of railway construction (which in all include some 60,000 bridges, nearly 1,000 tunnels and over 2,500 stations) surviving from before the year 1900. The dates quoted below generally apply to the completion of the work itself, but in a few cases where this is in doubt refer to the opening of the line as a whole.

THE TRAMWAY AGE

An embankment on Ryton Moor in Co: Durham is believed to date from the middle of the seventeenth century and to be the world's oldest 'railway' relic. An impressive earthwork nearby on Tanfield Moor is dated at 1725, and from the same system comes the famous but rather inaccessible 'Causey arch': built by Ralph Wood, this is a unique example of early-Georgian engineering. Other interesting survivals of this period are the Prestonpans, East Lothian, causeway (1722) and the Prior Park (Bath) incline (1731).

Good examples of later tramway engineering—sometimes marked by inclines and winding-houses—can be found near Chapel-en-le-Frith, Derbyshire (1796: the tunnel at Chapel Milton is probably Britain's oldest): Ashby-de-la-Zouch, Leicestershire (1802): Kilmarnock, Ayrshire (1811, with two good bridges): Penydarren, Glamorganshire (1812, again with two fine bridges): and the Collwng valley, Brecknockshire (1815). Near Robertstown, Glamorganshire, is the world's oldest metal bridge built for rail traffic (1811); and there are interesting tramway tunnels near Blaenavon, Monmouthshire (1815), at The Haie in the Forest of Dean (1810, and in use until very recently), at Glenfield, Leicestershire and Willsbridge, Gloucestershire (both 1832), and at Grosmont, Yorks (1836, by Robert Stephenson). On some of these lines—and on others, mentioned in Chapter Two—stone blocks and rarer items of furniture can be found *in situ*. Special mention should be made of the unique Haytor, Devonshire, granite tramway (1820).

Three lines which form links towards the railway age proper are the Surrey Iron (1805, with remains of its Godstone extension), the Stratford (1826, including John Rastrick's lovely bridge over the Avon), and the High Peak (1830). The last of these, by Josias Jessop, preserves numerous tunnels and inclines between Cromford and Whaley Bridge in Derbyshire, and affords throughout an outstanding example of tramway engineering.

Finally, two late tramway monuments difficult to classify are the section of the Ffestiniog railway up to Ddualt in Merionethshire (1836) and the grand Treffry viaduct in Cornwall (1847).

BRIDGES AND VIADUCTS

The oldest bridge of the railway age proper to survive in use (and largely in its original state) is Ignatius Bonomi's at Darlington

(1825): the fragment of an iron bridge from the same line preserved at York is also of technical as well as historic interest. On George Stephenson's Liverpool and Manchester railway (1830) the Sankey viaduct at Earlestown, Lancashire, and other bridges, established typically British forms which were to endure for over eighty years.

The least-rebuilt metal bridge on Robert Stephenson's London and Birmingham railway is the skewed one over Watling Street at Denbigh Hall, Buckinghamshire (1838), but numerous brick bridges remain in close to their original state. The short Ouse viaduct on this line is eclipsed by Joseph Locke's fine series of sandstone arches across the Weaver at Dutton in Cheshire (1837). From Brunel's Bristol railway come the daringly low brick bridges at Maidenhead, Berkshire (1839) and higher up the Thames and his characteristic Wharncliffe viaduct near Brentford, Middlesex (1838): from Francis Giles's Newcastle and Carlisle line the Wetheral, Cumberland, viaduct (also 1838): and from Grainger and Miller's Glasgow and Edinburgh that at Clifton Hall, East Lothian (1841). Another notable Scottish bridge can be found at Ballochmyle, Ayrshire (1848).

Probably the most elegant viaduct ever built is Rastrick and Mocatta's across the valley of the Sussex Ouse near Balcombe (1839): the same team were responsible for that carrying the Lewes branch just north of Brighton (1845). Other fine early viaducts can be seen at Penshaw, Co: Durham (1838), Stockport, Lancashire (1841), and by William Cubitt at Foord near Folkestone, Kent (1843) and Welwyn, Hertfordshire (1849). This latter year saw the opening of three great and contrasting works by Robert Stephenson—the 'high level' iron bridge at Newcastle upon Tyne, the Royal Border bridge of masonry at Berwick-upon-Tweed, and (on the far side of Britain) the tubular Britannia bridge over the Menai Strait to Anglesey. The last of these, now damaged, should be compared with its sister at Conway.

Brunel's Royal Albert bridge at Saltash outside Plymouth, opened in 1859, can be regarded as the last of the pioneering works of the railway age, though John Fowler's iron bridges in Shropshire (1863) are also worth study. Baker and Fowler's Forth bridge (1890) near Edinburgh stands for the era of mild steel, soon itself to be superseded by today's age of reinforced concrete.

TUNNELS AND CUTTINGS

Both the historic lines opened in 1830—the Canterbury and Whitstable and the Liverpool and Manchester—had lengthy tunnels. From the former that at Tyler Hill, Kent, survives disused and little changed: George Stephenson's Edge Hill tunnel at Liverpool is also substantially the original one, though the approach cuttings have been greatly widened.

On the first trunk lines, the outstanding such works are Robert Stephenson's tunnel at Kilsby, Northamptonshire (1838) and cuttings at Tring, Buckinghamshire (1837) and Roade, Northamptonshire, (1838) on the London and Birmingham railway: his father's Littleborough, Yorkshire, tunnel (1840) which was the first to pierce the Pennines: and Brunel's cutting at Sonning, Berkshire (1839) and series of tunnels down to Bath—of which that at Box, Wiltshire (1841) is easily the longest and most magnificently portalled. On Rastrick's Brighton line the Merstham, Surrey, tunnel and cuttings (1838) are particularly noteworthy. Joseph Locke normally eschewed tunnels, and his typical monument is hence the series of earthworks near Micheldever, Hampshire (1839): he was, however, responsible for the completion of the 'old' Woodhead tunnel in the Pennines (1845), three miles long and the world's first high mountain bore.

Numerous other tunnels are memorable for the engineering difficulties they presented; but they offer little to the sightseer except when adorned with unusual portals as at Clay Cross, Derbyshire (1840), Red Hill, Nottinghamshire (1840 again), Abbotscliff, Kent (1843), Audley End, Essex (1845), and Shugborough, Staffordshire (1847). The $4\frac{1}{2}$-mile Severn tunnel (1886), however, is unique not only for the dramas of its building but in remaining the world's longest underwater bore. This, too, forms a fitting postscript to the heroic age.

STATIONS

Although passenger bookings were accepted at the cottage at St John's Road, Stockton-on-Tees, in 1825, the title of 'Britain's first station' is normally reserved for John Foster's specially-constructed building at Liverpool Road, Manchester (1830). This may also be the world's first, though there is a challenger in Baltimore, Ohio.

Interesting early terminal stations survive—at least in part—at Birmingham, Curzon Street (Philip Hardwick, 1838), Brighton

(Mocatta, 1841), Bristol (Brunel, 1841) and elsewhere. The finest of later termini—using that term, where appropriate, to include matching hotels—are to be found in London, where the contrast of Lewis Cubitt's plain but noble King's Cross (1852) and Gilbert Scott's romantic fore-building to St Pancras (1876) is particularly striking. If a separate category be made of train sheds, the line of descent runs from Newcastle Central (John Dobson, 1850) through Paddington (Brunel, 1854) to St Pancras (Barlow, 1868) and so on to York (Thomas Prosser, 1877). Of detached hotels, the oldest survivor is Francis Thompson's at Derby (1840).

Noteworthy stations in various styles but of middling sizes are Carlisle Citadel (William Tite, 1847), Huddersfield (James Pritchett, 1847), Chester General (Thompson, 1848), Hull Paragon (G. T. Andrews, 1848), Monkwearmouth, Co: Durham (Dobson, 1848) and above all Newcastle Central (Prosser, 1855).

An outstanding series of minor stations, dating from about 1838 but unfortunately anonymous, can be found on the Newcastle and Carlisle railway. Also notable in the Tudor idiom are Tite's Windsor Riverside (1850), William Tress's Battle, Sussex (1852), and the small stations by Frederick Barnes in Suffolk, by Sancton Wood (probably) in Cambridgeshire, and anonymously in Staffordshire and Bedford-shire—all these dating from the late 1840s. Francis Thompson's stations such as Wingfield in Derbyshire (1841) and Holywell Junction in Flintshire (1848) deserve a study of their own. Darlington, North Road (1833) is a remarkable survivor; and it is hoped that all the stations mentioned on p. 123—and others—will have their admirers.

MISCELLANEOUS

Important miscellaneous works of railway engineering include George Stephenson's crossing of Chat Moss outside Manchester (1830) and Cubitt's of Whittlesey Mere, Northamptonshire (1849), Robert Stephenson's circular engine shed (now a theatre) at Camden Town, London (1837), and the railway towns of Locke's Crewe and Brunel's Swindon (both early 1840s). There are notable sea walls by Cubitt near the Shakespeare Cliff, Kent (1843), by Brunel at Dawlish, Devon (1845), and by Robert Stephenson at Penmaen, Caernarvon-shire (1849). But of all the handful of pioneering railway engineers it can still be said that he who seeks their monument need only look about him—on the 5,000 or more miles of their tracks.

Select Bibliography

As was mentioned in the introduction to this book, the railways of Britain have attracted an immense literature: even omitting that part of it which is concerned with locomotives and rolling stock this runs to many hundreds of volumes and thousands of articles, totalling tens of millions of words. The standard reference work, G. Ottley's *A Bibliography of British Railway History* (1965) is itself so massive that most readers will find more convenient E. T. Bryant's briefer but more critical *Railways—A Reader's Guide* (Bingley, 1968). A useful bibliographical appendix will also be found in Jack Simmon's *The Railways of Britain* (Macmillan, 2nd ed, 1968); and there even exists an American-based bibliography *of* railway bibliographies (Washington DC, 1938).

Those sampling this literature for the first time may find the present author's anthology, *The Railway-Lover's Companion* (Eyre & Spottiswoode, 1963), of some interest. The following notes concentrate on readily available books developing the aspects touched on in the present work—whose text has also mentioned a few more recondite primary sources.

The definitive volume on the tramway age—though largely in gazetteer form—is now Bertram Baxter's *Stone Blocks and Iron Rails* (David & Charles, 1966): unfortunately this has sins of both commission and omission. The transition to the railway period itself is covered in a well-researched recent volume by Michael Lewis somewhat misleadingly entitled *Early Wooden Railways* (Routledge & Kegan Paul, 1970); but the first main lines are still not adequately dealt with between one pair of covers.

For general histories of the classic railway age the reader can choose between the accounts—often well written and fascinating in their period detail, though sometimes superseded by later research—of such contemporaneous observers as Samuel Smiles, John Francis,

John Pendleton, Frederick Williams and William Acworth, and more modern and retrospective studies. Perhaps the standard such work, though one not quite irreproachable in its balance, is Hamilton Ellis's two-volume *British Railway History* (Allen & Unwin, 1954–59). The book by Jack Simmons mentioned above is uneven but compact and very good at its best, as is Michael Robbins's essayish *The Railway Age* (Routledge & Kegan Paul, 1962).

Every British railway company has at least one published history, these ranging from the officially sponsored accounts of the great post-1923 groups down to monographs on what Simmons terms 'minor and minimal' companies. These works (which normally include references back to primary sources) vary widely in their readability, logic and accuracy; and, as this is not the place to discuss a library which would itself run to at least fifty volumes, the reader is referred to the first paragraph of this survey. It should be mentioned, though, that David & Charles Ltd. is currently publishing a uniform series of railway histories, now edited by J. A. Patmore, based on the topographical regions of Britain rather than company structures. The arrangement is rather artificial and can hence become irritating; but the books themselves are admirably reliable and pay—as 'company' histories too rarely do—adequate attention to the industrial development of their various countrysides.

The information contained in the present book has largely been abstracted from such general histories, for it has already been mentioned that the civil engineering side forms one of the weakest departments of railway literature: little save technical studies of particular works has in fact appeared for well over a century, and certainly no overall modern history of the subject exists. Some useful little thematic guides have, however, been published by Ian Allan Ltd, these including *Tunnels* by Alan Blower and *Bridges* by David Walters. Further light is thrown by the biographies of the great railway engineers; for though L. T. C. Rolt has in the past complained that such men are under-appreciated he has not only done much to fill the gap with his definitive lives of *George and Robert Stephenson* (1960), *Isambard Kingdom Brunel* (1957) and others (Longman) and his *Victorian Engineering* (Lane, 1970), but has seen a number of other biographies newly written or reproduced from pious Victorian originals. Other works in this field include O. S. Nock's somewhat snippety *The Railway Engineers* (Batsford, 1955).

For railway architecture the situation is almost reversed. Probably through lack of adequate source material, virtually nothing has been published on the lives and works of such important figures as Mocatta and Thompson, and Lewis Cubitt does not even appear amid the dismal bishops of the *Dictionary of National Biography*. Something to remedy this lack is now being done by thesis-hungry students; but a full-length work on the great railway architects is much needed.

Station architecture has become a fashionable subject, though, and monographs on it have appeared almost as fast as distinguished stations have themselves been abandoned or destroyed. Christian Barman's well illustrated *An Introduction to Railway Architecture* (Art and Technics, 1950) viewed the theme with Festival-of-Britain optimism, and has been followed by David Lloyd and Donald Insall's *Railway Station Architecture* (David & Charles, n.d.) which deals with the visual aspects, by J. Horsley Denton's *British Railway Stations* (Ian Allan, 1964) which adopts a more practical approach, and by C. L. V. Meeks' international survey of *The Railway Station* (Yale University Press, 1957). Alan Jackson's *London's Termini* (David & Charles, 1969) is a fascinating guide to its subject. W. H. Chaloner has dealt with the development of a typical railway town in *The Social and Economic Development of Crewe* (Manchester University Press, 1951), and J. R. Kellett with *The Impact of Railways on Victorian Cities* (Routledge & Kegan Paul, 1969).

Finally, the present volume should be read in conjunction with other books in the Longman 'Industrial Archaeology' series, in particular John Snell's companion work on *Mechanical Engineering* and the forthcoming volume on *The Structural Use of Iron*.

Index